QE 899
.L 35

Late Mississippian and
Early Pennsylvanian Conodonts
Arkansas and Oklahoma

H. Richard Lane
Amoco Production Company
Tulsa, Oklahoma 74102

Joseph J. Straka II
Shell Canada Limited
Calgary, Alberta, Canada T2P2K3

THE
GEOLOGICAL SOCIETY
OF AMERICA

3300 Penrose Place · Boulder, Colorado 80301

SPECIAL PAPER 152

PUBLISHED BY
THE GEOLOGICAL SOCIETY OF AMERICA, INC.
3300 PENROSE PLACE
BOULDER, COLORADO 80301

Printed in The United States of America

*The printing of this volume has been made possible
through the bequest of
Richard Alexander Fullerton Penrose, Jr.,
and the generous support of all contributors
to the publication program.*

Contents

Acknowledgments

Gilbert Klapper of the University of Iowa served as our dissertations advisor. Discussions with Dr. Klapper on all aspects of this study have been invaluable in developing the taxonomic and biostratigraphic concepts presented herein.

W. M. Furnish, Brian F. Glenister, Holmes A. Semken, and Harrel Strimple of the University of Iowa; W. Bruce Saunders of Bryn Mawr University; G. A. Sanderson, H. J. Sullivan, and G. J. Verville of Amoco Production Company; J. H. Quinn of the University of Arkansas; L. R. Wilson of the University of Oklahoma, and M. K. Elias of the University of Oklahoma Research Institute all gave freely of their time and advice, to either or both of us, on various aspects of this study.

P. U. Rodda of the Bureau of Economic Geology at the University of Texas, J. E. Merida of the United States National Museum, P. K. Sutherland of the University of Oklahoma, and J. W. Huddle of the U.S. Geological Survey made type specimens in collections under their care available to us. R. L. Ethington of the University of Missouri, besides lending us type specimens, aided invaluably in resolving some taxonomic and nomenclatorial problems.

Lane is indebted to P. K. Sutherland, Thomas W. Henry, David Bowlby, T. L. Rowland, and Bruce N. Haugh, who were all associated with the University of Oklahoma in 1967, for generously providing descriptions of measured sections around Tenkiller Ferry Reservoir, Oklahoma, and for several days assistance while collecting samples, Later, P. K. Sutherland and Thomas W. Henry discussed many aspects of the stratigraphy of the Morrowan in Arkansas and Oklahoma. G. A. Massey of Amoco Production Company provided advice and assistance in the laboratory. Lane is particularly indebted to Amoco Production Company for providing field expenses and laboratory and drafting facilities.

Straka is indebted to M. R. Thomasson, Head of Geologic Research, Shell Development Company, Houston, for suggesting his topic, and to the management of the Denver area Shell Oil Company for providing field expenses. Innumerable members of the staff of the Oklahoma division of Shell Oil Company gave freely of their time and talents to aid Straka in this investigation. Special thanks are extended to Robert Oswald, Donald Beard, Robert Forester, Jack Mase, and H.

R. Downs of Shell Oil Company for criticisms and suggestions relating to this study.

We acknowledge Shell Oil Company for contributing toward the printing costs of this investigation.

Charles Collinson, M. Avcin, D. L. Dunn, J. W. Huddle, and M. Gordon have critically read the manuscript.

Abstract

The type area for the Springerian stratigraphic sequence is in southern Oklahoma (Carter County) and that of the Morrowan is in northwestern Arkansas (Washington County). In the past, both series have been employed as comprising successive time-stratigraphic serial subdivisions of the North American Lower Pennsylvanian. A conodont zonal scheme developed herein provides a means for correlations between both type sections and demonstrates that the Springerian overlaps with the underlying Chesterian and overlying Morrowan. Therefore, usage of the Springerian as a viable subdivision of the Lower Pennsylvanian should be discontinued.

In the type Springerian, the upper part of the Goddard Formation and Rod Club Member of the Springer Formation correlate with the Menard through the Grove Church sequence in the Illinois Basin. The "B" Shale Member contains an undiagnostic Upper Mississippian fauna, and the Target Limestone Lentil of the Lake Ardmore Member contains conodonts that correlate with the Cane Hill Member of the Hale Formation of the type Morrowan. The Lake Ardmore through Primrose sequence correlates with the Hale through Woolsey Member of the Bloyd Formation in Arkansas. The Dye Shale and Kessler Members of the Bloyd contain two upper Morrowan conodont zones, and the position of the Trace Creek Shale Member with respect to the Morrowan-Derryan (=Atokan) Boundary cannot be precisely determined. Morrowan conodont collections from northeastern Oklahoma can be correlated with the type Morrowan zonal scheme.

The zonal scheme developed in both areas is present in Morrowan rocks in west Texas and Nevada. One taxon is newly described.

1

Introduction

The Mississippian-Pennsylvanian Boundary in most areas of the North American midcontinent is easily distinguished due to an unconformity of considerable magnitude separating the two systems. However, it has been recognized for some time that the Springer Formation in the Ardmore Basin of southern Oklahoma bridges this unconformity (Goldston, 1922; Moore, 1934; Moore and others, 1944; Weller and others, 1948; Tomlinson and McBee, 1959).

In the past, the Springerian Series (Cheney and others, 1945; and modified by Elias, 1956a) has been treated as the lowest subdivision of the standard midcontinent Pennsylvanian sequence (Tomlinson and McBee, 1962). In its type area on the southern flanks of the Arbuckle Mountains, the Springerian Series comprised all of the Noble Ranch Group (Straka, 1972), excluding the Goddard Formation. The Morrowan Series (Harlton, 1938; and modified by Dott, 1941) has its type area in Washington County, Arkansas. The Morrowan is composed of the Hale and Bloyd Formations in the type area, and it has been envisioned as succeeding the Springerian in the standard midcontinent Lower Pennsylvanian (Cheney and others, 1945; Moore and Thompson, 1949).

In our opinion, it has never been conclusively demonstrated that the Springerian and Morrowan Series are nonoverlapping time-stratigraphic units. Furthermore, one school of thought contends that the Springerian is, at least in part, Chesterian (Upper Mississippian) in age and, therefore, should not be employed as the lowest subdivision of the Pennsylvanian System (Laudon, 1958; Branson, 1959; Furnish and others, 1964). The general uncertainty surrounding the age of the Springerian probably reflects the lack of adequate faunal and floral control in the type area.

In the last decade, uppermost Mississippian and Lower Pennsylvanian conodonts have come under intensive study. The most significant studies in terms of conodont form-taxonomy and biostratigraphic succession include Rexroad and Burton (1961), Dunn (1965, 1966, 1970a, 1970b), Lane (1967), Koike (1967), Higgins and Bouckaert (1968), Webster (1969), Meischner (1970), Lane and others (1971, 1972), and Straka (1972).

The objective of this study is to present a comprehensive systematic treatment of biostratigraphically important conodonts from the Late Mississippian and Early

Pennsylvanian rocks exposed in the type Springerian and Morrowan regions. The areas of investigation are depicted on Figure 1. This systematic treatment allows us to document a conodont faunal zonation that can be utilized in biostratigraphic correlations between the types Chesterian, Morrowan, and Springerian. This zonal scheme also has general applicability throughout the western United States. We feel that this study demonstrates that the Springerian Series is not a viable time-stratigraphic subdivision of the Lower Pennsylvanian.

This investigation represents the combination of two dissertations completed at the University of Iowa—Lane's on the type Morrowan and Straka's on the type Springerian. The joint publication of our information avoids unnecessary duplication and presents a more coherent and compact study of these important Upper Mississippian and Lower Pennsylvanian conodont sequences.

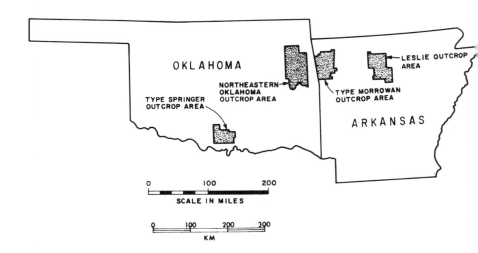

Figure 1. Generalized map showing areas of investigation.

Stratigraphy

SOUTHERN OKLAHOMA, CARTER COUNTY

Because numerous studies have been published concerning stratigraphic units adjacent to and comprising the Springerian Series in its type area, we have summarized previous authors' stratigraphic nomenclature on Figure 2. It is readily apparent that this sequence of rocks has suffered from nomenclatorial instability throughout its history of investigation. A complete discussion of each study would be lengthy and would serve no real purpose for our intentions herein. Consequently, the reader is referred to the original works listed along the top of Figure 2 for each author's justification of his scheme of stratigraphic nomenclature. The stratigraphic nomenclature used in this study is the same as Straka's (1972), and that study should be treated as a supplement to the current investigation.

In constructing Figure 2, it has been necessary in several cases to interpret the original authors' intentions in the absence of data and an adequate explanation. One important interpretation on our part is discussed below.

Goldston (1922) illustrated several cross sections of strata above the Sycamore Limestone in the northern Ardmore Basin. He depicted a sequence of six or seven sandstone units alternating with shales below the first occurrence of a limestone within his newly described Springer Member of the Glenn Formation. In an attempt to determine which sandstone units Goldston represented, we transferred his lines of profile to several recent geologic maps. In addition, each area of Goldston's sections was visited to establish field identification of exposed sandstone units. In all cases, it was found that the basal sandstone of Goldston's Springer Member could be only the lowest of the three Lake Ardmore Sandstones. Goldston's conclusion that the base of the Springer Member occurs 400 ft (121.9 m) below the ammonoid-bearing sandstone (=Primrose Sandstone) at the Santa Fe Railroad cut (Locality 3, herein) would approximately correspond with the base of the Lake Ardmore Sandstone of current usage. Furthermore, we believe the 130 ft (39.6-m) sandy shale below the Primrose ammonoid horizon is the shaly eastward extension of the middle and upper sandstone units of the Lake Ardmore. Thus, Goldston almost certainly regarded the Lake Ardmore Sandstone of current usage as the basal unit of his Springer Member. Most subsequent authors believed that Goldston intended the Rod Club to be the basal unit of his Springer Member.

The locations of our sampled sections in southern Oklahoma are illustrated on Figure 3.

Noble Ranch Group

Straka (1972) proposed the Noble Ranch Group to include those strata from the top of the Caney Shale to the base of the Primrose Member of the Golf Course Formation (for example, the Goddard and Springer Formations). The term "Noble Ranch" is derived from a ranch on the Caddo anticline where most of the succession is exposed; however, the type section was designated as north and east of Springer, Oklahoma, in secs. 1–6, T. 3 S., R. 2 E., Carter County, Oklahoma, where a nearly complete composite sequence of these strata is exposed in several north-south-flowing stream valleys. Reference sections for the group are close by on the Caddo anticline, on the Goddard ranch, east of Gene Autry, Oklahoma, and in Redoak Hollow, north of Milo, Oklahoma.

Goddard Formation

The Goddard Formation of this report corresponds precisely to Westheimer's (1956) type Goddard Formation and to the Goddard Formation of Cheney and others (1945), Moore and Thompson (1949), Harlton (1956), Elias (1956a, 1960), Tomlinson and McBee (1959), Peace (1965), and Lang (1966). The stratigraphic boundaries of the formation as delimited by Westheimer (1956, p. 393) are the top of the Mississippian Caney and the base of his (Tomlinson's, 1929) Springer Formation. The upper and lower boundaries are generally distinct, but in the absence of the Rod Club Sandstone in outcrop (base of Tomlinson's Springer Formation), the contact is extremely difficult to delimit. The gray to gray green, soft Goddard shales are identical in outcrop appearance and x-ray analysis (Weaver, 1958) to those Springer shales overlying the Rod Club.

Redoak Hollow Member

Elias (1956a, p. 83–85) named this unit, and described its lithology and fauna. We measured 70 ft (21.3 m) exposed interbedded sandstone and shale strata at Elias' fossil locality IV in Redoak

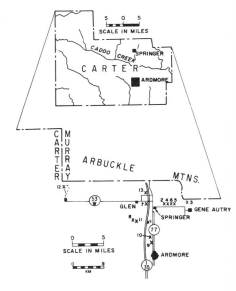

Figure 3. Localities collected and described in the type Springerian area, Carter County, Oklahoma.

Hollow, but consider this a low estimate of the actual thickness because of the weak resistance of the unit to weathering and consequent development of cover.

Tiff Member

Tomlinson (1959, p. 323) originally named this member of the Goddard Formation the Grindstone Creek Member, for exposures along Grindstone Creek at NE1/4SE1/4SE1/4SE1/4 sec. 16, T. 3 S., R. 1 E., Carter County, and stated that it occurred "about 500 feet below Rod Club Member" (of Springer Formation). However, Tomlinson and Bennison (1960, p. 123) recalled that subsequent to publication of the name "Grindstone Creek," Tomlinson learned that the term was preoccupied by a member of the Millsap Lake Formation of Texas. Tomlinson proposed a new name "Tiff Member" by means of a rubber stamp overprint on reprints of his article (Tomlinson, 1959), and the name was registered with the Geologic Names Committee of the U.S. Geological Survey.

The type locality of this unit is now inundated by a stock pond. This section (Locality 11) was measured and channel collected for conodonts before water encroachment. The rock sequence exposed at Locality 7 was also measured, and channel samples were taken. The stratigraphic relations of the strata at Locality 7 to superjacent and underlying beds are unclear. However, it is apparent from physical correlation of this section with the type Tiff section that both units are the same, and faunal assemblages from each locality confirm this correlation. Locality 7 should be considered the prime reference section for the Tiff Member of the Goddard Formation.

Springer Formation

The Springer was first called a formation in a stratigraphic study by Tomlinson (1929). His Springer Formation included, however, strata that later workers, and this report, exclude from the top of the formation (for example, the Primrose Sandstone and Gene Autry Shale). Tomlinson drew the base of the Springer Formation at the base of the Rod Club Sandstone, as have most later authors.

Moore and Thompson (1949) were the first to apply the term "Springer Formation" in the same sense as adopted herein. These authors' Springer Formation was underlain by the Goddard Formation, with the Rod Club Sandstone as the basal Springer unit, and was overlain unconformably by the Primrose Sandstone, which was the basal unit of their Morrowan Stage. Later authors who followed this stratigraphic interpretation and restriction of Tomlinson's original definition include Harlton (1956), Westheimer (1956), Elias (1956a, 1960), Tomlinson and McBee (1959), Peace (1965), and Lang (1966). These authors differ among themselves, however, regarding grouping of the Springer strata with the Goddard Formation, age assignment of the Springer sequence, and rank of the term "Springer."

In accordance with the consensus, our concept of the Springer Formation is that proposed by Moore and Thompson (1949). The formation comprises three sandstone members, which are, in ascending order, the Rod Club Member at the base of the formation; the "Overbrook" Member; and the Lake Ardmore Member (which includes the Target Limestone Lentil). The sandstone members are separated by shale units, herein designated "C" Shale, "B" Shale, and Academy Church Shale Member. The upper boundary of the Springer Formation is marked by the base of the overlying Primrose Member of the Golf Course Formation. The Springer Formation straddles the Mississippian-Pennsylvanian Boundary as indicated by the contained conodont faunas.

Rod Club Member

The lithology and occurrence in outcrop of the basal member of the Springer Formation is adequately described and known from past literature (Westheimer, 1956; Tomlinson, 1929, 1959; Tomlinson and McBee, 1959; and others).

"C" Shale Member

The informally designated "C" Shale lies above the Rod Club Sandstone. Its presence is known from subsurface and from swale topographic expression. No true exposures of this shale are known to us from the region north of Ardmore, Oklahoma, and none are documented in the literature. Tomlinson (1959, p. 322–323) described three localities where he believed this shale was exposed. All three localities were visited by Straka. Two sites south of Ardmore expose strata in uncertain stratigraphic settings; consequently, they were not sampled. Strata at the third locality (Locality 7) north of Ardmore also display uncertain stratigraphic relations; however, they are distinctive lithologically and, therefore, were sampled. The strata at Locality 7 correlate with the Tiff Member of the Goddard Formation, in contrast to Tomlinson's determination.

"Overbrook" Member

Lang (1966, p. 14) reported subsurface studies that indicated the type Overbrook (designated by Tomlinson, 1929) is equivalent to the so-called Goddard Sandstone of the subsurface. Therefore, this unit was not believed to be the same sand as the so-called Overbrook Member of the Springer Formation on the Caddo anticline. If this interpretation is correct, perhaps the term "Overbrook" should be discontinued as a member of the Springer Formation and a substitute name proposed with its type section on the Caddo anticline. The sandstone called Overbrook on the Caddo anticline is that which has been regarded as typical and constitutes the "Overbrook concept" in the literature, but in light of Lang's surface and subsurface findings, it has been shown to be nonequivalent with the type Overbrook.

The "Overbrook" Sandstone Member of this report is used in the sense of that middle sandstone unit of the Springer exposed on the Caddo anticline, particularly along the southern edge of the Noble Ranch, north of Ardmore. This unit is prominently exposed here forming part of the dams of City Lake and Lake Ardmore.

Samples of this sandstone at Localities 5, 6, and 8 would not respond to either acetic acid or Stoddard solvent digestion; consequently, no conodonts were retrieved.

"B" Shale Member

The informally designated "B" Shale crops out above the "Overbrook" Sandstone. Member. This shale was found exposed east of Springer, Oklahoma, at the new Locality 4. No other exposures of this shale are known in the area of investigation. The "B" Shale is like the "C" Shale and Goddard Formation lithologically and cannot be distinguished from them in outcrop except where the underlying "Overbrook" or overlying Lake Ardmore Sandstone are also exposed. The lower 135 ft (41.15 m) of this shale are exposed at the above locality and were sampled. The remaining 350 ft (106.7 m) to the base of the Lake Ardmore are covered and remain unsampled.

Lake Ardmore Member

At its type locality on the Caddo anticline, the Lake Ardmore Member is a single sandstone unit approximately 15 ft (4.6 m) thick. The type locality of Tomlinson and McBee's (1959) Lake Ardmore Formation has a different connotation because their type locality is found along the south flank of the Arbuckles, where the member is comprised of three sandstone units with intervening shale sections. Those authors' Lake Ardmore Formation comprised the strata from the base of the first sandstone above the "B" Shale to the base of the Primrose. They stated that the sandstone at the type locality of the Lake Ardmore Member (on the Caddo anticline) possibly corresponds to the middle sandstone unit of their Lake Ardmore Formation (Tomlinson and McBee, 1959, p. 11).

Tomlinson and McBee's type locality should serve only as a reference locality for the Lake Ardmore Member. A more fully developed section of Lake Ardmore Sandstone is exposed in that reference section than at the type locality on the Caddo anticline.

Target Limestone Lentil of the Lake Ardmore Member

The Target Limestone is not well exposed anywhere along its outcrop extent. The best exposure is at the type locality established by Bennison (1954, p. 913). We could not find either the top or bottom of this lentil exposed anywhere along the outcrop belt, but estimated a thickness of 15 to 20 ft (4.6 to 6.1 m) based on the topographic ridge supported by the vertically standing beds of this unit.

The Target lies between the basal and middle sandstone units of the Lake Ardmore Member, about 50 ft (15.2 m) above the lowermost, according to Tomlinson and McBee (1959, p. 12). The limited extent of surface exposure and absence in the subsurface of this limestone unit prompted Straka (1972) to consider it of minimal stratigraphic rank.

Academy Church Shale Member

Straka (1972) proposed this shale member for the uppermost unit of the Springer Formation. It has an apparent conformable relation to the overlying Primrose Sandstone Member of the Golf Course Formation at the type locality. In the subsurface, however, the Academy Church Shale is commonly truncated or completely removed by pre-Primrose erosion.

A complete surface exposure of this shale is at Locality 3 where the succession of strata from approximately the middle of the Lake Ardmore Member of the Springer Formation through the Gene Autry Shale Member of the Golf Course Formation is continuously exposed.

Visually, in outcrop, and mineralogically (Weaver, 1958), this shale unit differs from the other underlying shales in the Springer Formation ("B" and "C") and Goddard Shale.

It is also faunally different from those underlying shales. On these bases, Straka (1972) distinguished this Springer shale unit nomenclatorially from those below it, and proposed the term "Academy Church Shale Member." The 250 ft (76.2 m) thick shale unit is generally dark gray in color and blocky, with interbedded, black, brittle shale beds 2 to 4 ft thick. The type locality is at E1/2NE1/4NW1/4SE1/4 and W1/2NW1/4NE1/4SE1/4 sec. 1, T. 3 S., R. 2 E., Carter County, Oklahoma, where the member is continuously exposed in a southward-flowing tributary from Cool Creek, along the west side of the Gulf, Colorado, and Santa Fe Railroad embankment. The name is derived from a wooden frame church which stands on the Primrose Sandstone ridge just west of the type locality.

Golf Course Formation

This report is concerned only with the Primrose Member, the basal unit of the Golf Course Formation, as that formation was defined by Harlton (1956, p. 138–139).

Primrose Member

The Primrose Sandstone Member is easily distinguished from the Springer Sandstone Members in outcrop by its coarser grain size, glauconite content, and calcareous nature. For these reasons, and by recognition of an unconformity at the base of the unit in the subsurface, the Primrose has been classified by several past authors (Cheney and others, 1945; Moore and Thompson, 1949; Elias, 1956a; Harlton, 1956; Tomlinson and McBee, 1959) as distinct from the Springer. Earlier workers, however, did include the Primrose with the Springer (Goldston, 1922; Tomlinson, 1929; Harlton, 1934).

The Primrose was measured and sampled at three localities (1, 2, 3) and a correlation chart of the member and underlying and overlying beds also exposed at these localities is presented by Straka (1972, Fig. 3). We interpret the "lower Primrose" of Elias (1956a) at the Caddo Village locality (Locality 1) to be the lower portion of the Primrose Sandstone Member that is "shaled-up," contrary to Elias' belief (1956a, p. 98; 1969, oral commun.) that it is an earlier development of Primrose strata not represented elsewhere.

NORTHERN ARKANSAS

The locations of our sampled sections in northwestern Arkansas are depicted on Figure 4. For a complete history of investigation of the northern Arkansas Section, see Henbest (1953) and Lane (1967).

Pitkin Limestone

The Pitkin Limestone was named by Ulrich (1904) for limestone exposures near the post office of Pitkin in Washington County, Arkansas. The unit overlies the Fayetteville Shale and underlies the Cane Hill Member of the Hale Formation in Washington and Crawford Counties, Arkansas. In north-central Arkansas, the Upper Mississippian Imo Formation overlies the Pitkin. Henbest (1962a) designated the type section as being in the 30- to 50-ft (9.1- to 15.2-m) cliff along the West Fork of the White River at the base of Bloyd Mountain (center of the west side of sec. 4, T. 14 N., R. 30 W., Washington County, Arkansas). The formation consists of a bluish gray, oölitic to micritic, fossiliferous limestone, which is medium to massively bedded. A complete section of the Pitkin was not measured for this report. To the east, in Van Buren County, the unit is certainly much thicker, and at Peyton Creek (Locality 25) the uppermost 15 ft (4.6 m) of the unit consists of a medium-bedded, dark gray, phosphatic, oölitic limestone that is interbedded with dark gray shale which may attain a maximum thickness of 2 ft (0.61 m) between individual limestone beds. Easton (1942) and Gordon (1965, p. 31) have given more complete discussions of the Pitkin Limestone, and the reader is referred to those works for further information.

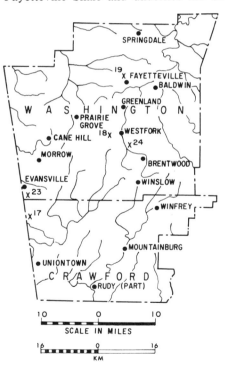

Figure 4. Localities collected in the type Morrowan area, Washington and Crawford Counties, Arkansas.

As noted by Lane (1967, p. 923), the

top of the Pitkin Limestone in northwestern Arkansas and northeastern Oklahoma yields conodonts which belong in the *Kladognathus-Cavusgnathus naviculus* Zone, whereas in north-central Arkansas the top of the Pitkin is younger and bears conodonts indicative of the *Adetognathus unicornis* Zone. This heterochroneity of the top of the Pitkin in both areas is probably due to postdepositional erosion in northwestern Arkansas and northeastern Oklahoma.

Imo Formation

Gordon (1965) named the Imo Formation for exposures of approximately 350 ft (106.7 m) of dark gray to black clay shale that contains several massive beds of fine-grained calcareous sandstone and fossiliferous calcareous shales in the lower two-thirds of the formation. The formation has its type section in Sulphur Springs Hollow, a tributary of Bear Creek in the SE1/4SE1/4 sec. 3, and the NW1/4NW1/4 sec. 11, T. 13 N., R. 17 W., Searcy County, Arkansas.

Furnish and others (1964) reported that a goniatite fauna consisting of the genera *Delepinoceras* Miller and Furnish, *Anthracoceras* Frech, *Cravenoceras* Bisat and *Eumorphoceras* Girty in the Imo indicates its correlation with the uppermost *Eumorphoceras* Zone, lower Namurian (E2) of Europe. A small conodont fauna which was recently retrieved from the calcareous sandstone beds near the base of the formation at Peyton Creek includes *Cavusgnathus naviculus*, *Gnathodus bilineatus*, and *G. commutatus commutatus*. This fauna is Upper Mississippian in age, and due to the known occurrence of the *Adetognathus unicornis* Zone in the underlying Pitkin Limestone, these conodonts can confidently be assigned to the *A. unicornis* or *R. muricatus* Zones. The age of the Imo above the cephalopod horizon at Peyton Creek is not known.

Hale Formation

Henbest (1953) recognized that the Hale Formation could be divided into two mappable units in western Arkansas, the shale below and the sandstone above. He designated the units the Cane Hill and Prairie Grove Members, respectively.

Cane Hill Member

The Cane Hill Member is extremely variable in thickness (45 ft or 13.7 m maximum) and is generally restricted to western Arkansas. This member consists of dark brown to black splintery shales and dark brown fucoidal sandstones. A few stringers of limestone are known locally, and commonly the unit has a basal conglomerate. Henbest (1962a, p. D39) designated the type section of the Cane Hill to be at the Evansville Mountain roadcut (= Locality 23).

We recovered several small conodont faunas from the Cane Hill which belong in the *R. primus* Zone (Lower Pennsylvanian). However, at several sections (Localities 17 and 18) the *R. primus* Zone is developed in rocks that are indistinguishable from the Prairie Grove Member. Therefore, in places, the base of the Prairie Grove Member is a lateral equivalent of the Cane Hill, and in one case

(Locality 18) rocks assigned by us to the Prairie Grove but belonging in the *R. primus* Zone rest unconformably on the Pitkin Limestone.

Prairie Grove Member

Lane (1967, p. 924) summarized the pertinent information about the Prairie Grove Member, and the reader is referred to that study. However, the approximate additional 20 ft (6.1 m) of unmeasured Prairie Grove at the type section that is mentioned by Lane (1967) is in place and is actually about 70 ft (21.3 m) thick. Therefore, at Evansville Mountain (Locality 23, herein) the Prairie Grove is approximately 158 ft (47.9 m) thick.

Bloyd Formation

The Bloyd Formation was originally named by Purdue (1907) and Henbest (1962a, p. D40), and its type section was designated as being in the southwestern part of Bloyd Mountain extending from the center, on the north side of sec. 4, T. 14 N., R. 30 W., Washington County, Arkansas. The Bloyd consists of fossiliferous limestones, shales, and sandstones with one thin (1 ft, 0.3 m, or less) coal seam (Baldwin coal) about in its center. Currently, five members of the Bloyd Formation (in its type area) are recognized and these are, in ascending order: Brentwood Limestone, Woolsey Shale, Dye Shale, Kessler Limestone, and Trace Creek Shale. Information about these members is given below.

Brentwood Limestone Member

Lane (1967, p. 924) summarized pertinent information about the Brentwood Limestone Member, and the reader is referred to that study.

The top of the Brentwood Limestone becomes younger to the west at the expense of the overlying Woolsey Member. At Bloyd Mountain (near Locality 24) the Woolsey is 45 ft (13.7 m) thick (Henbest, 1953, p. 1943). At our Locality 24 on Bloyd Mountain, the complete Brentwood and lower part of the Woolsey is exposed. There, the upper part of the Brentwood belongs in the *Neognathodus bassleri bassleri* Zone, and the lower Woolsey yielded conodonts indicative of the *Idiognathodus sinuosis* Zone. To the west at Evansville Mountain (Locality 23), the Woolsey is no more than 2.5 ft (0.76 m) thick, and the conodonts in the upper part of the underlying Brentwood belong in the *I. sinuosis* Zone. Therefore, based on these two sections, the Woolsey Shale thins westward as the top of the Brentwood becomes younger.

Woolsey Shale Member

Lane (1967, p. 924) summarized pertinent information about the Woolsey Shale. However, it now appears that the Woolsey Shale thins westward to nearly a feather edge near the Arkansas-Oklahoma border from its maximum development on Bloyd Mountain.

Dye Shale Member

Henbest (1962b) described and named the Dye Shale Member for exposures on Bloyd Mountain E1/2 sec. 3, to the center of the north side of sec. 4, T. 14 N., R. 30 W., Washington County, Arkansas. The member was named for rocks that lie above the Woolsey and below the Kessler Limestone Members of the Bloyd Formation. The member varies from 60 to 110 ft (18.3 to 33.5 m) in thickness (Henbest, 1962b) and consists of dark gray to black shales with a few thin limestone stringers. Of importance to this report is the caprock of the Baldwin coal that occurs at the base of the Dye Shale Member. The caprock is from 0 to 25 ft (0 to 7.6 m) thick and generally consists of a dark brown to black quartz pebble conglomerate; calcareous sandstones and fossiliferous limestones are commonly developed. At Evansville Mountain (Locality 23), the caprock is 5.5 ft (1.7 m) thick and consists of a chocolate brown siliceous sandstone, which grades upward to a highly fossiliferous limestone. A thin band of the Baldwin coal, no more than 1 in. (2.5 cm) thick, is exposed immediately below the caprock at this locality. The shale overlying the caprock and underlying the Kessler Limestone Member at Evansville Mountain is poorly exposed and was not measured or described.

Kessler Limestone Member

The Kessler Limestone Member was named by Simonds (1891, p. 103) for exposures "in 16 N., 30 W." Henbest (1962a) pointed out that Kessler Mountain is in R. 31 W., not R. 30 W., and that two limestones occur "high up on the slope" of Kessler Mountain. The lower and less well developed limestone represents the caprock of the Baldwin coal. Henbest (1962a) clarified the situation by establishing the algal-foraminiferal and commonly oölitic Kessler Limestone exposed near the center of the SE1/4 sec. 25, T. 16 N., R. 31 W., as the type locality.

Figure 5. Localities collected in the Morrowan outcrop belt of Mayes, Cherokee, and Muskogee Counties, Oklahoma.

In this report, the Kessler Limestone Member was measured and sampled at Evansville Mountain (Locality 23). The member is exposed near the top of the mountain in the first major road cut as State Highway 59 descends the southern slope. At this locality, the Kessler is 11.5 ft (3.5 m) thick and consists of a gray, oölitic to algal limestone. The unit is rubbly near its base, and individual beds thicken and thin laterally.

Trace Creek Shale Member

Henbest (1962b) named the Trace Creek Shale Member of the Bloyd Formation for rocks that lie between the top of the Kessler Limestone Member of the Bloyd Formation and the base of the Greenland Sandstone Member of the Winslow Formation. The member was named for exposures on Bloyd Mountain near Trace Creek. The type locality is the same as that for the Bloyd Formation, on the southwestern side of Bloyd Mountain from the center of the E1/2 sec. 3, to the center of the north side of sec. 4, T. 14 N., R. 30 W., Washington County, Arkansas. The Trace Creek Shale Member consists dominantly of a brittle black shale with a few thin conglomeratic, sandy, and (or) calcareous beds distributed throughout. Henbest (1962b) stated that the Trace Creek Shale is typically 60 to 70 ft (18.2 to 21.3 m) thick. At Evansville Mountain, only the lower 12.5 ft (3.8 m) were measured, but it is estimated that the total thickness is greater than 50 ft (15.2 m). Conodont faunas retrieved from the Trace Creek are poorly understood; consequently, the position of the member with respect to the Morrowan-Derryan Boundary is not known.

NORTHEASTERN OKLAHOMA

The locations sampled in northeastern Oklahoma are depicted on Figure 5.

In northeastern Oklahoma, the upper part of the Hale Formation and lower part of the Bloyd Formation are difficult to distinguish. In many cases, we have separated the formations based on our faunal data. Therefore, our usage of these two formations in northeastern Oklahoma must be considered as being in a provisional manner.

A discussion of the stratigraphy of the Morrowan in northeastern Oklahoma is not presented because a comprehensive study encompassing the sections investigated is being prepared by P. K. Sutherland and T. W. Henry at the University of Oklahoma. That study should be used as a supplement for interpretation of our data from northeastern Oklahoma. Our correlations of the sections studied in northwestern Arkansas with those in the northern part of the study area in northeastern Oklahoma are illustrated on Figure 6. Correlation of sections in the vicinity of Tenkiller Ferry Reservoir in northeastern Oklahoma is illustrated on Figure 7.

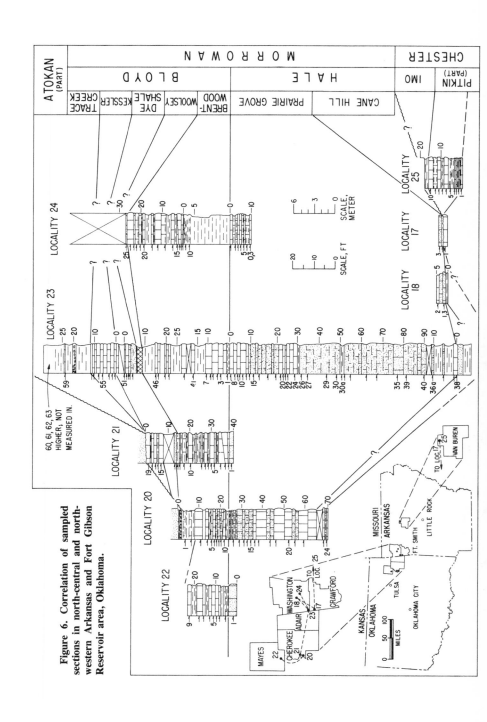

Figure 6. Correlation of sampled sections in north-central and north-western Arkansas and Fort Gibson Reservoir area, Oklahoma.

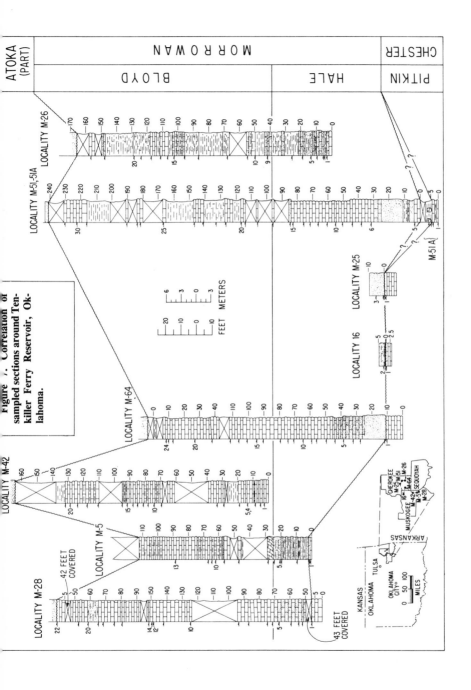

Figure 7. Correlation of sampled sections around Tenkiller Ferry Reservoir, Oklahoma.

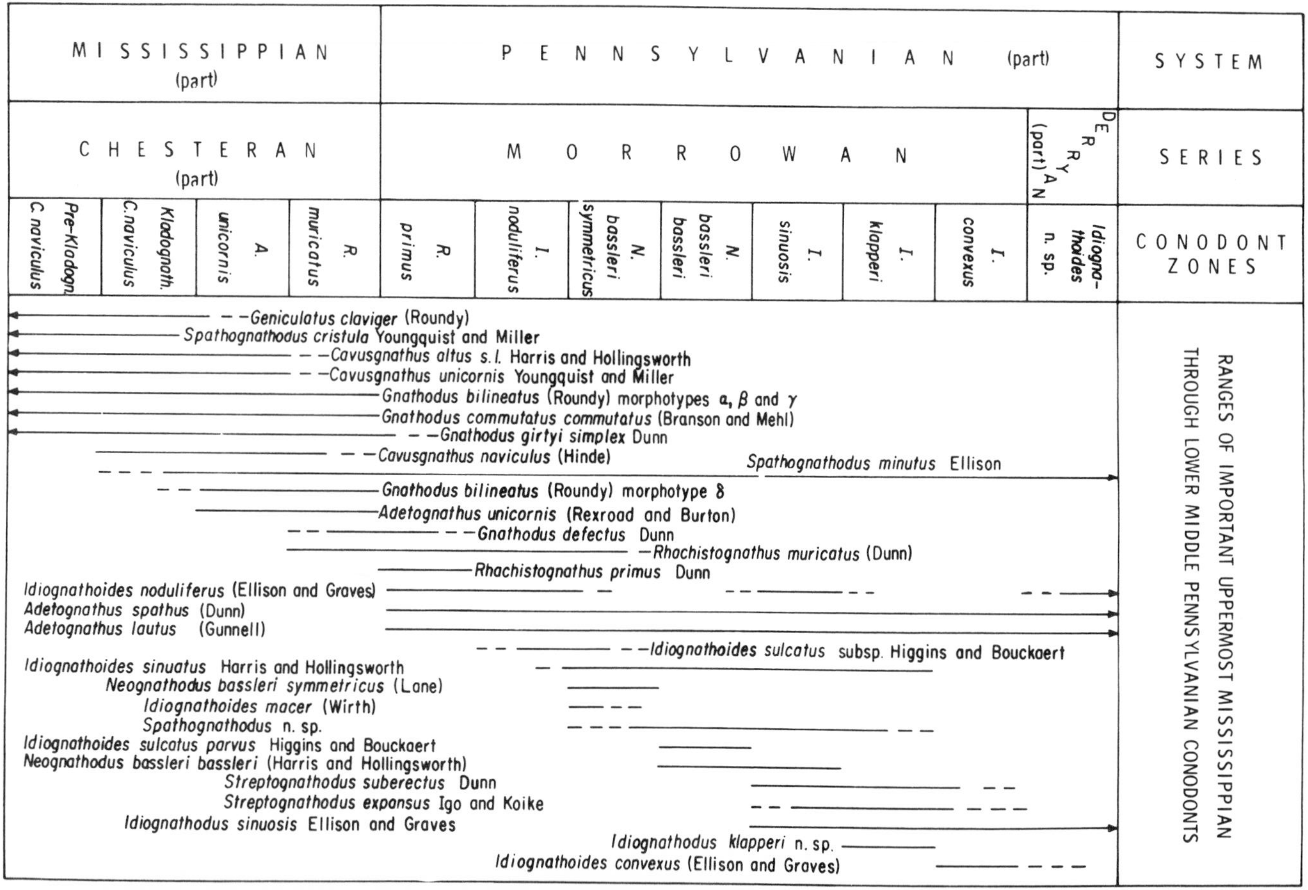

Figure 8. Ranges of important uppermost Mississippian through lower Middle Pennsylvanian conodonts.

Conodont Zonation

Our zonal scheme for this report is depicted along with the ranges of strati-graphically important conodonts, on Figure 8. The relation of this zonation to other conodont zonal schemes proposed for this interval is presented on Figure 9.

MISSISSIPPIAN

Basically, two distinctive conodont faunal intervals can be recognized in the Mississippian stratigraphic units investigated for this study. The *Kladognathus-Cavusgnathus naviculus* Zone of Collinson and others (1971, p. 387) occurs in the portion of the Goddard Formation below the Tiff Member and in at least the uppermost few feet of the Pitkin Limestone in northwestern Arkansas and northeastern Oklahoma. *Cavusgnathus naviculus* has not been found to occur in the basal beds (Redoak Hollow Member and lower) of the Goddard Formation, and only an undifferentiated lower and middle Chesterian age can be assigned to that interval based on its conodont fauna. *Kladognathus primus* has not been found in any of our samples, but the range of *C. naviculus* below the appearance of *Adetognathus unicornis* is sufficient to delineate the zone. In the Illinois Basin, the *Kladognathus-Cavusgnathus naviculus* Zone occurs from the base of the Walche Limestone Member of Menard Formation to the base of the Grove Church Shale (Collinson and others, 1971, p. 387).

The *Adetognathus unicornis* Zone succeeds the *Kladognathus-Cavusgnathus naviculus* Zone and was originally defined as the range of the name-bearer (Collinson and others, 1962) based on its occurrence in the Grove Church Shale in the Illinois Basin. The full range of *A. unicornis* is not developed in the Illinois Basin because of the Mississippian-Pennsylvanian unconformity at the top of the Grove Church Shale (Collinson and others, 1971, p. 387). In southern Nevada, a more complete sequence is developed across the Mississippian-Pennsylvanian Boundary (Webster, 1969; Dunn, 1970b). There, *A. unicornis* occurs in the lower part of the Indian Springs Formation and is associated with *Rhachistognathus muricatus* in the upper part of the Indian Springs and lower part of the Bird Spring Formations (Webster,

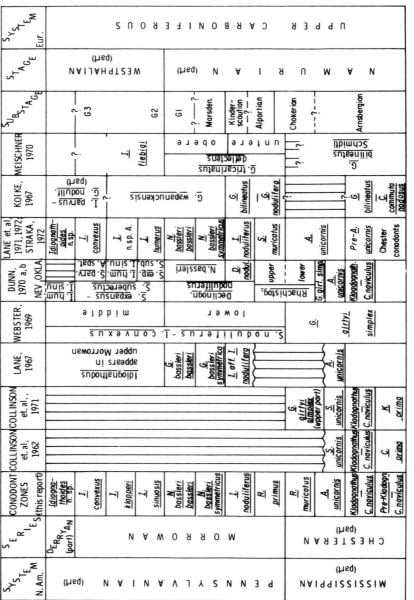

Figure 9. Interpretation of the relation of various Upper Mississippian and Lower Pennsylvanian conodont zonal schemes and sequences in North America, Japan, and Europe. Studies from which information was derived are indicated at the top of each

1969; Dunn, 1970a; Lane and others, 1972). *Rhachistognathus muricatus* is not known to occur in Upper Mississippian rocks of the Illinois Basin. We believe that the interval in Nevada in which *A. unicornis* and *R. muricatus* are associated represents a new faunal zone that is not present in the Illinois Basin due to the unconformity at the top of the Mississippian. Therefore, we find it necessary to redefine the top of the *A. unicornis* Zone so that it corresponds with the appearance of *R. muricatus* and to propose a new uppermost Mississippian zone above the *A. unicornis* Zone. A discussion of both zones is given below.

Adetognathus unicornis Zone

The *A. unicornis* Zone is defined as the range of the name-bearer below the first occurrence of *Rhachistognathus muricatus*. Conodonts which occur in the zone include *A. unicornis, Cavusgnathus naviculus, C. unicornis, C. altus, Gnathodus bilineatus* (all morphotypes), *Hibbardella milleri,* and in Nevada, *Gnathodus defectus* and *Gnathodus girtyi simplex.* The zonal name-bearer first appears slightly below the base of the Tiff Member of the Goddard Formation and ranges at least to the middle of the Rod Club Member of the Springer Formation in southern Oklahoma. Except for a very sparse conodont fauna in the lower part of the "B" Shale of the Springer Formation, the interval from the middle part of the Rod Club to the base of the Target Limestone Lentil was barren of conodonts. Thus, the upper range of *A. unicornis* in the type Springerian area is not established. The *A. unicornis* Zone is not present in northeastern Oklahoma and northwestern Arkansas due to an unconformity at the top of the Pitkin Limestone. However, the zone is present in the upper part of the Pitkin Limestone of north-central Arkansas. Although *A. unicornis* has not been found in the Imo Formation of north-central Arkansas, a small fragmentary fauna containing *C. naviculus* has been found in the calcareous sandstone below the main ammonoid-producing horizon at Peyton Creek.

The lower part of the "B" Shale of the Springer Formation in southern Oklahoma yielded a fauna consisting entirely of *Gnathodus girtyi intermedius.* Although this form corresponds closely with *Gnathodus girtyi simplex,* its stratigraphic significance cannot be adequately analyzed due to the total absence of other conodonts. However, *G. girtyi intermedius* has previously been reported only from Mississippian strata (Globensky, 1967).

Rhachistognathus muricatus Zone

The *Rhachistognathus muricatus* Zone is defined as the range of the name-bearer below the first occurrence of the lowermost Pennsylvanian *Rhachistognathus primus* or *Adetognathus lautus,* or both. This interval is equivalent to the lower subzone of the *Rhachistognathus* Zone of Dunn (1970b, p. 2967). Other conodonts that characterize the *Rhachistognathus muricatus* Zone include *Cavusgnathus naviculus, Gnathodus bilineatus, G. commutatus commutatus, G. girtyi simplex, Gnathodus defectus,* and *Adetognathus unicornis.*

The zone is known to occur in the upper part of the Indian Springs Formation and lower part of the Bird Spring Formation in southern Nevada, based on range information presented by Webster (1969), Dunn (1970a, 1970b), and Lane and others (1972). The zone is absent in the Illinois Basin, northwestern Arkansas, and northeastern Oklahoma due to the Mississippian-Pennsylvanian unconformity. However, the time-stratigraphic interval may be present in the Imo Formation in north-central Arkansas, and "C" Shale, "Overbrook," and "B" Shale Members of the Springer Formation in southern Oklahoma.

The *Spathognathodus muricatus* faunal unit of Lane and others (1971, 1972) is younger than the *Rhachistognathus muricatus* Zone and conforms with the Lower Pennsylvanian *R. primus* Zone discussed later.

Gnathodus girtyi simplex ranges from at least the top of the lower Chesterian Battleship Wash Formation to the top of the Mississippian in southern Nevada (Webster, 1969; Lane and others, 1972). Lane and others (1972) found the form-subspecies in the base of the Lower Pennsylvanian La Tuna Formation in west Texas. Therefore, the *G. girtyi simplex* Zone of Webster (1969) has broad limits and cannot be utilized in a refined zonal scheme. Dunn's (1970b) usage of the *G. girtyi simplex* Zone is different than its original definition and is based on incomplete range information. Therefore, the *G. girtyi simplex* Zone of Dunn (1970b) is not a viable biostratigraphic unit.

PENNSYLVANIAN

Lane and Straka (Lane and others, 1971) outlined a conodont faunal unit scheme for the type Morrowan and Pennsylvanian portion of the type Springerian based on the information presented herein. The following scheme is very similar to that of Lane and Straka, but important changes are discussed below.

Rhachistognathus primus Zone

The zone is defined as the range of the name-bearer and is lowest Morrowan in age. The zone corresponds with the upper subzone of Dunn's (1970b) *Rhachistognathus* Zone. Other conodonts which occur in the zone include *Adetognathus lautus, Idiognathoides noduliferus, Rhachistognathus muricatus, Adetognathus spathus*, and *I. sulcatus*.

The zone occurs in the Cane Hill Member of the Hale Formation in northwestern Arkansas and the Target Limestone Lentil in southern Oklahoma. Thus, the Target Limestone correlates with the Cane Hill, and the Mississippian-Pennsylvanian Boundary falls within the Springerian Series of Elias (1956a). The zone is also known to occur in the lower part of the Bird Spring Formation in southern Nevada (Dunn, 1970b; Lane and others, 1972) and in the lower part of the La Tuna Limestone of West Texas (Dunn, 1966; Lane and others, 1972).

The *Spathognathodus muricatus* faunal unit of Lane and Straka (Lane and others, 1971, 1972) corresponds with the *R. primus* Zone herein.

Idiognathoides noduliferus Zone

This zone is characterized by *Idiognathoides noduliferus* above the last occurrence of *Rhachistognathus primus* and below the first occurrence of *Neognathodus bassleri symmetricus*. Other conodonts that occur in the zone include *Adetognathus lautus, Idiognathoides sulcatus*, and, near the top, *Idiognathoides sinuatus* appears. The zone occurs in the lower part of the Prairie Grove Member of the Hale Formation in Washington County, Arkansas. This interval also occurs in the Lake Ardmore through the Academy Church Shale Members of the Springer Formation and the basal part of the Primrose Member of the Golf Course Formation, Carter County, Oklahoma.

The zone is not known to occur in the lower part of the Bird Spring Formation at Arrow Canyon, southern Nevada, based on information presented by Webster (1969) and Lane and others (1972). However, the occurrence of the form-species in samples SN–30, 31 of Dunn at the Lee Canyon section, southern Nevada, may correspond with the zonal interval.

Neognathodus bassleri symmetricus Zone

The zone is defined as the range of the name-bearer below the first occurrence of *N. bassleri bassleri*. Other conodonts that characterize the fauna include *Adetognathus lautus, Idiognathoides sinuatus, Spathognathodus* n. sp. A., *Ozarkodina delicatula, Spathognathodus minutus*, and, in southern Nevada, *Rhachistognathus muricatus*.

Neognathodus bassleri symmetricus appears in the upper part of the Prairie Grove Member of the Hale Formation and ranges into the lower part of the Brentwood Member of the Bloyd Formation in northwestern Arkansas. The form-subspecies occurs in the upper part of the Hale Formation and lower part of the Bloyd Formation in northeastern Oklahoma, and near the base of the Primrose Member of the Golf Course Formation in southern Oklahoma. Lane (1967) reported an impoverished fauna comprised dominantly of *N. bassleri symmetricus* near the base of the Brentwood in northwestern Arkansas, but that interval is restricted to sections studied in Washington County, Arkansas. The zone is known to occur from 209 to 225 ft (63.7 to 68.6 m) above the base of the La Tuna Limestone in west Texas and at 126 ft (38.4 m) above the base of the Bird Spring Formation at Arrow Canyon, southern Nevada (Lane and others, 1972).

Neognathodus bassleri bassleri Zone

The zone corresponds with the ranges of *N. bassleri bassleri* and *Idiognathoides sulcatus parvus*, below the first appearance of the form-genera *Idiognathodus* and *Streptognathodus*. Other conodonts which occur in the zone include *Adetognathus lautus, Idiognathoides sinuatus*, and, in southern Nevada, *Rhachistognathus muricatus*.

The zone occurs in the upper part of the Brentwood Limestone Member of the Bloyd Formation at Locality 24 in northwestern Arkansas. *N. bassleri bassleri* occurs in the middle and upper part of the Primrose Member of the Golf Course Formation, Carter County, Oklahoma, and with *I. sulcatus parvus* in the lower part of the Bloyd Formation in northeastern Oklahoma. Lane and others (1972) reported the zone to occur from 256 to 270 ft (78 to 82.3 m) above the base of the La Tuna Formation, Helms Peak section, west Texas, and probably at 149 ft (45.4 m) above the base of the Bird Spring Formation, Arrow Canyon section, southern Nevada.

Idiognathodus sinuosis Zone

The zone is defined as the overlap in ranges of the name-bearer with *N. bassleri bassleri*. Other conodonts which occur in the zone include *Idiognathoides sinuatus, Adetognathus lautus, Streptognathodus expansus,* and *Spathognathodus minutus.*

The zone is known to occur in the base of the Woolsey Member of the Bloyd Formation in central Washington County, Arkansas, and in the top of the Brentwood in the western part of the county. Also, the zone occurs within the Bloyd Formation of northeastern Oklahoma and in the upper part of the Primrose Member of the Golf Course Formation, Carter County, Oklahoma.

This zone corresponds with the *Idiognathodus humerus* faunal unit of Lane and Straka (Lane and others, 1971, 1972). The new taxonomic assignment of the zonal name-bearer to the element-pair species *I. sinuosis* necessitates a change in zonal nomenclature.

Idiognathodus klapperi Zone

The caprock of the Baldwin coal (lower part of the Dye Shale Member) yields a conodont fauna containing *Idiognathodus klapperi*. This form-species is only known from the lower part of the Dye Shale Member of the Bloyd Formation at Locality 23. However, rocks above the caprock of the Baldwin coal and below the Kessler Limestone Member of the Bloyd Formation in northwestern Arkansas have not been sampled. Consequently, the upper range of *I. klapperi* is not known. Other conodonts that occur in the unit include *Adetognathus lautus, Idiognathoides sinuatus,* and *Idiognathodus sinuosis*.

This zone corresponds with the *Idiognathodus* n. sp. A. faunal unit of Lane and Straka (Lane and others, 1971, 1972).

Idiognathoides convexus Zone

Idiognathoides convexus (Ellison and Graves) appears at the base of the Kessler Limestone Member of the Bloyd Formation in Washington County, Arkansas, and the upper limit of the range of the element-pair species is not known. However, the zone is tentatively defined as the range of the name-bearer below the appearance in the lower Derryan of *Idiognathoides* n. sp. (Lane and others, 1972). The zonal definition must remain in a preliminary manner until the relation of the ranges of *I. convexus* and *Idiognathoides* n. sp. is better understood. Undiagnostic faunal elements of the *I. convexus* Zone include *Idiognathodus sinuosis* and *Adetognathus lautus*.

In northern Arkansas, the zone is restricted to the Kessler Limestone Member, but further study and collecting may show that zone extends down into the underlying Dye Shale and up into the overlying Trace Creek Shale.

Trace Creek Shale Member

The Trace Creek Shale Member of the Bloyd Formation yields a fauna containing *Idiognathodus sinuosis, Adetognathus gigantus, Adetognathus lautus,* and *Idiognathoides sinuatus.* This fauna indicates only a post-Brentwood, Pennsylvanian age. Further collections and study of the conodonts of the Trace Creek Shale Member are necessary to clarify the position of this member with respect to the Morrowan-Derryan Boundary as discussed by Lane and others (1972).

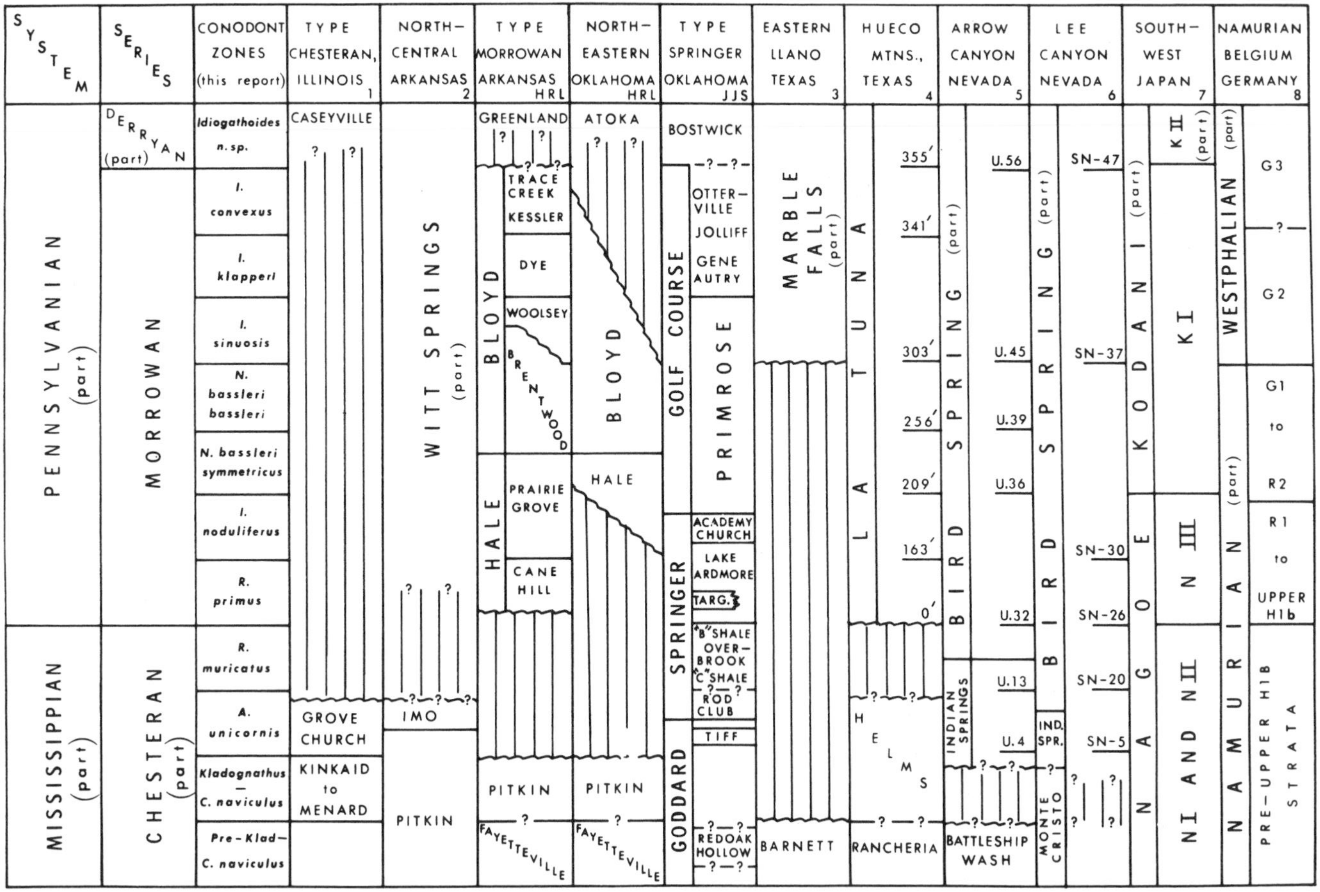

Figure 10. Correlation of important stratigraphic sequences in North America, Japan, and Europe based on their conodont faunas. The letters HRL and JJS at the top of certain columns refer to information derived from H. R. Lane's and J. J. Straka's dissertations, respectively. Information in numbered columns was derived from the following investigations: 1. Collinson and others, 1971; 2. Gordon, 1965; Lane, 1967; 3. Dunn, 1970a, 1970b; 4. Lane and others, 1972; 5. Webster, 1969; Lane and others, 1972; 6. Dunn, 1970a, 1970b; 7. Koike, 1967; 8. Higgins and Bouckaert,

Conodont Biostratigraphy

NORTH AMERICA

Mississippian-Pennsylvanian Boundary

Straka (1972) has discussed and illustrated correlations between the type Springerian and type Morrowan areas. Those correlations are based on information presented herein, and the reader is referred to Straka's study and our Figure 10 for our correlations between the two areas.

Dunn (1970b) and Straka (1972) discussed the problem of placement of the Mississippian-Pennsylvanian Boundary in the type Springerian. Dunn provisionally placed it at the base of the Lake Ardmore Member of the Springer Formation, whereas Straka placed it at the base of the Target Limestone Lentil of the Lake Ardmore Member of the Springer Formation. Straka (1972) found that approximately the lower one-third of the "B" Shale Member of the Springer Formation contained *Gnathodus girtyi intermedius* Globensky, and assigned that fauna to the Mississippian because of that subspecies occurrence in definite Mississippian age rocks in the Windsor Group of the Atlantic provinces of Canada (Globensky, 1967). However, Globensky's occurrences of that subspecies seem to be much older than Straka's because of its association with other form-species (that is, *Apatognathus? geminus*, *A.? porcatus*, and *Spathognathodus scitulus*) that can be dated as Meramec or lower Chester. Thus, Straka's material may be homeomorphic with Globensky's subspecies. In any case, the forms reported as *Gnathodus girtyi intermedius* by Straka from the lower one-third of the "B" Shale and described and illustrated herein (see Fig. 33: 1-10, 24, 26, 27) have affinities with the Mississippian form-species *Gnathodus girtyi* Hass and are assigned to the Mississippian. There are no known exposures of the approximate upper two-thirds of the "B" Shale. Consequently, we provisionally place the Mississippian-Pennsylvanian Boundary near the base of the overlying Lake Ardmore Member of the Springer Formation. Dunn (1970b) also placed the boundary at the same position. The Target Limestone Lentil, a local carbonate unit in the lower part of the Lake Ardmore Member, carries conodonts indicative of the *Rhachistognathus primus* Zone and thus is lowermost Morrowan in age.

In northwestern Arkansas (Washington and Crawford Counties) the Cane Hill Member of the Hale Formation rests unconformably on the Pitkin Limestone. The Pitkin Limestone in that area yields conodonts indicative of the *Kladognathus-Cavusgnathus naviculus* Zone. The overlying Cane Hill yields conodonts that we assign to the *Rhachistognathus primus* Zone and is, by definition, lowermost Pennsylvanian in age. Therefore, in the type Morrowan region, the Mississippian-Pennsylvanian unconformity represents a time interval equivalent to the *Adetognathus unicornis* and *Rhachistognathus muricatus* Zones, and probably parts of the *Kladognathus-Cavusgnathus naviculus* and *Rhachistognathus primus* Zones. In north-central Arkansas (Van Buren County), the Imo Formation (Gordon, 1965) overlies the Pitkin Limestone. There (Locality 25) the upper part of the Pitkin Limestone (and probably the lower part of the Imo Formation) yields conodonts that belong to the *Adetognathus unicornis* Zone (Lane, 1967). Lower Pennsylvanian rocks in that area have not been sampled for conodonts, and therefore the extent of the unconformity is not known. In any case, uppermost Mississippian rocks are younger in north-central Arkansas than in the northwestern part of the state.

In northeastern Oklahoma, the Hale Formation overlies the Pitkin Limestone. The upper part of the Pitkin Limestone in that area carries conodonts indicative of the *Kladognathus-Cavusgnathus naviculus* Zone, and the lower part of the overlying Hale Formation yields conodonts belonging to the *Idiognathoides noduliferus* Zone. Therefore, all of the *Adetognathus unicornis, Rhachistognathus muricatus,* and *R. primus* Zones, and probably parts of the *Kladognathus-Cavusgnathus naviculus* and *I. noduliferus* Zones, are missing. In one of the most westerly sections (Locality M-5) the *Neognathodus bassleri symmetricus* Zone rests on a conodont fauna in the top of the Pitkin, which can probably be assigned to the *Kladognathus-Cavusgnathus naviculus* Zone. The specimen from the upper Pitkin at that locality, which is assigned to *R. muricatus,* is poorly preserved and was probably derived from the conglomeratic limestone at the contact of the Pitkin and Hale.

The above information suggests that the Mississippian-Pennsylvanian unconformity in northern Arkansas and northeastern Oklahoma becomes more extensive in a westerly direction. Although this may be the case, our density of sample locations in the area of investigation is not sufficient to exclude the possibility of an undulating unconformable surface, where the top of the Mississippian and the base of the Pennsylvanian vary in age due to paleotopography.

Webster and Lane (1967, p. 516) and Webster (1969) placed the Mississippian-Pennsylvanian Boundary at the top of the *Rhipidomella nevadensis* brachiopod Zone (Sadlick, 1955) in southern Nevada. Webster (1969) stated that the top of the *R. nevadensis* Zone in southern Nevada corresponds with the contact between his *Gnathodus girtyi simplex* and *Streptognathodus noduliferus-Idiognathoides convexus* Assemblage Zones. We have found that *Gnathodus girtyi simplex* ranges into lowest Pennsylvanian strata (at least through the *R. primus* Zone) in southern Nevada and west Texas. At Arrow Canyon, the highest occurrence of *R. nevadensis* is in Webster's (1969, Fig. 5) Unit 26, whereas the last occurrence of *G. girtyi simplex* in our collections is in Unit 32. Webster (1969, Fig. 11) reported the first occurrence of the name-bearers of his lowest Pennsylvanian *Streptognathodus noduliferus-Idiognathoides convexus* Assemblage Zone in Unit 32. Thus, the limits of Webster's two zones overlap and do not conform with the top of the *Rhipidomella nevadensis* Zone in Arrow Canyon. Collinson and others (1971, p. 388) stated that the upper part of Webster's Mississippian *G. girtyi simplex* Assemblage Zone is younger than the *Adetognathus unicornis* Zone in the Illinois Basin and Mississippi Valley area. In light of our range information on *G. girtyi simplex,* and the works of Dunn (1970a, 1970b) and Lane and others (1972), we believe that the upper part of Webster's highest Mississippian conodont zone (the portion discussed by Collinson and others, 1971, p. 388) actually conforms with the *Rhachistognathus muricatus* Zone (=lower subzone of the *Rhachistognathus* Zone; Dunn, 1970b) and the lower Pennsylvanian *R. primus* Zone (=upper subzone of the *Rhachistognathus* Zone; Dunn, 1970b).

Lane and others (1972) placed the Mississippian-Pennsylvanian Boundary at the first appearance of *Adetognathus lautus* and *A. gigantus* (= *A. lautus,* herein) based on collections from Arkansas, Texas, and Nevada. Their *S. muricatus* faunal unit is the same as the *Rhachistognathus primus* Zone in the current study.

Dunn (1970a, 1970b) studied many sections in the western United States which span the Mississippian-Pennsylvanian Boundary. He placed the boundary at the top of the lower subzone of his *Rhachistognathus* Zone or at the appearance of *Rhachistognathus primus.* This horizon corresponds with the top of the *Rhipidomella nevadensis* brachiopod Zone (Sadlick, 1955).

Dunn (1970a) recorded the joint occurrence of *Neognathodus bassleri, Idiognathodus sinuosis,* and *Rhachistognathus primus* in his basal samples from the Marble Falls Limestone of the eastern Llano area of central Texas. We attribute the presence of *R. primus* in the faunas to reworking. If Dunn's data and our interpretation of reworking are correct, then the base of the Marble Falls at his localities 1 and 2 (Dunn, 1970a, p. 318-319) correlates with the *Idiognathodus sinuosis* Zone of the present study. Thus, the base of the Marble Falls Limestone correlates with the Woolsey Member of the Bloyd Formation in northwestern Arkansas, and the Mississippian-Pennsylvanian unconformity in the eastern Llano area of Texas is much more extensive that Dunn (1970b, Fig. 2) indicated.

Dunn (1970b) delineated three biostratigraphic intervals in the Upper Mississippian above the *Kladognathus-Cavusgnathus naviculus* Zone. These are (1) the *Adetognathus unicornis* Zone, defined as the range of the name-bearer below the first occurrence of *G. girtyi simplex;* (2) the *G. girtyi simplex* Zone, defined as the range of the name-bearer below the first occurrence of *Rhachistognathus muricatus;* and (3) the lower subzone of the *Rhachistognathus* Zone, defined as the range of *Rhachistognathus muricatus* below the first occurrence of *R. primus.* Webster (1969, Figs. 11-15) and Lane and others (1972) found that in southern Nevada *G. girtyi simplex* ranges down into the Battleship Wash Formation. Thus, Dunn's (1970b, Fig. 4) range for *G. girtyi simplex* is incomplete. Consequently, his *Gnathodus girtyi simplex* Zone cannot be recognized as a viable zonal interval.

In the current study, we place the Mississippian-Pennsylvanian Boundary at the contact between the *Rhachistognathus muricatus* and *R. primus* Zones (=contact between the lower and upper subzones of the *Rhachistognathus* Zone of Dunn, 1970b). Although both zones do not occur sequentially in either the type Chesterian or type Morrowan, we feel that the appearance of *A. lautus* and *R. primus* is a distinctive change in the faunas, which can be easily utilized for differentiating the two series. Furthermore, this horizon conforms with the top of the *Rhipidomella nevadensis* brachiopod Zone in southern Nevada and has been conventionally recognized as the position of the boundary in that area. Until some sort of international agreement on placement of the boundary is put forth, we will continue to use the above scheme of definition.

Morrowan

Lane (1967) proposed the first zonal scheme for Morrowan rocks based on collections from the type Morrowan of northwestern Arkansas. In that study, his *Idiognathoides* aff. *I. noduliferus* Zone was found to occur in the lower part of the Prairie Grove Member of the Hale Formation. The *Gnathodus bassleri symmetricus* Zone succeeded the *I.* aff. *I. noduliferus* Zone and was found to occur in the upper part of the Prairie Grove Member of the Hale Formation and the lower part of the Brentwood Member of the Bloyd Formation. The overlying *G. bassleri bassleri* Zone was found in the upper part of the Brentwood Member. The Woolsey Member of the Bloyd Formation yielded the first representatives of the genus *Idiognathodus* and thus could be easily distinguished from underlying strata based on its conodont fauna.

Webster (1969, p. 23) proposed one zone for what he thought to be lower and middle Morrowan rocks of southern Nevada. This interval, termed the *Streptognathodus noduliferus* (= *Idiognathoides noduliferus*, herein)- *Idiognathoides convexus* Assemblage Zone, could be divided into three parts: a lower subzone containing abundant *Rhachistognathus muricatus* in addition to the name-bearers; an upper subzone containing abundant *Gnathodus bassleri* and numerous *Idiognathodus delicatus;* and a middle subzone lacking either of the supplemental elements of the lower and upper subzones. Lane and others (1972) found that *Rhachistognathus muricatus* ranges as high as the *Neognathodus bassleri bassleri* Zone in southern Nevada. Thus, Webster's lower subzone is lower to middle Morrowan in age. His middle subzone is probably upper Morrowan, and the upper part of his middle subzone corresponds with the lower Derryan *Idiognathoides* n. sp. Zone of Lane and others (1972). Webster (1969, Pl. 5, figs. 9, 14, 15; and in Lane and others, 1971, Pl. 1, fig. 26), Dunn (1970a, Pl. 64, fig. 14), and Merrill (in Lane and others, 1971, Pl. 1, figs. 29, 30) have figured forms which they referred to *Gnathodus bassleri.* We do not believe that these conodonts are consubspecific with *Neognathodus bassleri symmetricus* or *N. bassleri bassleri,* but are representatives of the Derryan form-species *Neognathodus colombiensis* (Stibane). Furthermore, the interval yielding *N. colombiensis* in southern Nevada contains Foraminifera and fusulinids indicative of Mamet and Skipp's (1970) lower Derryan Zone 21 (see Lane and others, 1972). Thus, the upper subzone of Webster's (1969) *Streptognathodus noduliferus-Idiognathoides convexus* Zone, which is characterized by occurrences of *N. colombiensis,* is Derryan in age. In light of the above evidence, Webster's *Streptognathodus noduliferus-Idiognathoides convexus* Assemblage Zone encompasses all the Morrowan and the lower part of the Derryan and corresponds only in a very small part with the *Idiognathoides noduliferus* Zone of Lane and others (1971, 1972), Straka (1972), and as presented herein.

Dunn (1970b) proposed four zones for the Morrowan of Nevada and Utah and five zones for the Morrowan of Muskogee and Sequoyah Counties, Oklahoma, based on data he presented earlier (Dunn, 1970a). The only difference between the two zonal schemes is the addition of the *Neognathodus bassleri* Zone in Oklahoma which Dunn considered to be equivalent to the upper part of his *Declinognathodus noduliferus* (= *Idiognathoides noduliferus*) Zone in Nevada and Utah. We agree with this correlation, but Lane and others (1972) demonstrated that the *Neognathodus bassleri symmetricus* Zone and conodonts indicative of the *Neognathodus bassleri bassleri* Zone are present below the first occurrence of *Streptognathodus suberectus* at Arrow Canyon in southern Nevada. Therefore, a more precise correlation of this interval with the type Morrowan stratigraphic intervals is now possible. Dunn (1970b) recognized three zones above his *Declinognathodus noduliferus* Zone in southern Nevada and *Neognathodus bassleri* Zone in Oklahoma. These are, in ascending order, (1) the *Streptognathodus expansus-S. suberectus* Zone, the base defined at the appearance of the name-bearers; (2) the *Idiognathodus humerus-I. sinuosis* Zone, the base defined at the appearance of the name-bearers; and (3) the *Streptognathodus parvus-Adetognathus spathus* Zone, the base defined at the uppermost occurrences of *Idiognathodus humerus* and *I. sinuosis* and the top by the uppermost occurrence of *Adetognathus spathus*.

It should be noted that Dunn's (1970b, p. 2968) recognition of the above zonal scheme as applicable to strata in northeastern Oklahoma is based on faunas from two localities in that area. One spot sample was taken by Dunn (1970a, p. 318) at his locality 4, and it yielded *Streptognathodus expansus* and the element-pair species *I. sinuosis*. Thus, this spot sample may be assigned to his *I. humerus-I. sinuosis* Zone. At Dunn's (1970a, p. 318) locality 3, none of the elements that characterize his *S. expansus-S. suberectus* or *I. humerus-I. sinuosis* Zones were recovered by Dunn. *Streptognathodus parvus* occurred in only sample 33 at that locality, and *Adetognathus spathus* was lacking. Thus, in northeastern Oklahoma it is difficult to recognize Dunn's highest three zones (Dunn, 1970b, p. 2968) based on his presented data (Dunn, 1970a).

In southern Nevada, Lane and others (1972) demonstrated that Zone 21 of Mamet and Skipp (1970) is well developed in Arrow Canyon in the approximate 500 ft (152.4 m) of Bird Spring strata below the first occurrence of *Profusulinella*. Mamet and Skipp (1970) assign this zone to the early Atokan (=Derryan), and Lane and others (1972) recovered Zone 21 fusulinids in the lower 15 ft of the type Derryan section in New Mexico. Zone 21 forams and fusulinids have never been recovered from the type Morrowan section. In Lee Canyon, Nevada, Rich (1961, p. 1166) recovered Zone 21 fusulinids as low as his sample 44 (=sample 62 of Dunn, 1970a, p. 315, 322), and *Profusulinella* was recovered in his sample 109. Rich (1961, p. 1166, 1168) questionably assigned this interval to the Morrowan but stated that he realized the interval might be, in part at least, correlative with the lower Atokan. Later, Dunn (1970a, p. 315) assigned Rich's samples 44–85 to the Morrowan using *Profusulinella* as the lowest indicator of Derryan (Atokan) rocks. Certainly *Profusulinella* can be used to distinguish Morrowan from Derryan rocks, but *Profusulinella* does not appear at the base of the Derryan Series (Lane and others, 1972). Lane and others (1972) suggested that strata as low as Dunn's sample SN–47 at Lee Canyon are Derryan in age based on their conodont faunas, and the SN–47 horizon conforms with the lower limit of Dunn's *I. humerus-I. sinuosis* Zone. Thus, strata above the base of the latter zone in Lee Canyon are probably Derryan in age.

The forms that Merrill (Lane and others, 1971) refers to *Gnathodus bassleri symmetricus* and *G. bassleri bassleri* (Pl. 1, figs. 29, 30; Fig. 4) are conspecific with *Neognathodus colombiensis*. Therefore, the Boggs and Lower Mercer Formations are Derryan in age, and his report of *G. noduliferus* from the Lowellville may correspond with known Derryan occurrences of that form-species.

EXTRA-NORTH AMERICA

Igo and Koike (1964, 1965) reported Carboniferous conodont faunas from the Omi and Akiyoshi Limestones of central and southwestern Japan, respectively. The fauna from the Omi Limestone is at least as young as, and probably correlates with, the *Idiognathodus sinuosis* Zone herein, and therefore is middle to upper Morrowan in age. The fauna from the Akiyoshi Limestone cannot be precisely dated at this time, but it is suggestive of a middle Morrowan age.

Koike (1967) reported a succession of conodont faunas from the Nagoe and Kodani Formations in southwestern Japan which range in age from Upper Mississippian to Middle Pennsylvanian in age. His uppermost Mississippian *Gnathodus bilineatus-G. noduliferus* (= *Idiognathoides noduliferus*, herein) Zone is puzzling in that we have never found those two species to occur together, except in cases where *G. bilineatus* has clearly been reworked into faunas containing *I. noduliferus*. We would suggest that this interval is probably Morrowan in age and corresponds approximately with our *Idiognathoides noduliferus* Zone. However, it is possible that the form which Koike (1967, Pl. 3, fig. 10) refers to *G. noduliferus* actually is *G. girtyi simplex*, and Koike's NIII interval may then be Upper Mississippian in age. In general, Koike's illustrations do not permit one to make precise taxonomic judgments. The overlying *Gnathodus wapanuckensis* (= *Neognathodus bassleri*) Zone should approximately correspond with our *Neognathodus bassleri symmetricus* through *I. convexus* Zone. The specimens illustrated by Koike on Plate 1, figures 22 and 24, from KI strata probably do belong in *Neognathodus bassleri*. However, his figured specimens on Plate 1, figures 23 and 25, from lower KII strata probably belong in *Streptognathodus parvus* and are approximately upper Morrowan in age. The joint appearance of *Gnathodus* cf. *G. roundyi* (= *Neognathodus* n. sp., Lane and others, 1972), *Pseudostaffella*, and *Eoschubertella* in the lower part of Koike's *Idiognathodus parvus-Gnathodus noduliferus* Zone at about 180 ft (54.7 m) above the base of his generalized columnar section (Koike, 1967, Table 3) conforms with the base of the lower Derryan Zone 21 of Mamet and Skipp (1970). Therefore, Koike's Morrowan-Atokan Boundary should be lowered to conform with the first appearance of the above-mentioned forms (Figs. 9, 10).

Higgins and Bouckaert (1968) studied conodont faunas from the Namurian of Belgium. Their data suggest that the Mississippian-Pennsylvanian Boundary should fall within the upper part of the *Homoceras beyrichianum* Zone, Chokerian (H1), in western Europe. This suggestion is based on the appearance of *Rhachistognathus muricatus* and *Idiognathoides noduliferus* in the upper part of the zone at Higgins and Bouckaert's localities at Ronet (locality VIII), Blaton (locality XX), and Bolland (locality I). Dunn (1970b, p. 3965) placed the boundary somewhere in H1. Further-

ore, the appearance of form-species *Idiognathoides corrugatus* and *I. attenuatus*
t the base of R2 would suggest that the base of the *N. bassleri symmetricus*
one in North America conforms with the base of the Marsdenian (R2) in western
urope. The form which Higgins and Bouckaert (1968, p. 18, Pl. 5, fig. 9) report
om the base of R1 is not conspecific with *Idiognathoides corrugatus*. Later,
ouckaert and Higgins (1970) suggested that the base of the Kinderscoutian Stage
rrelates with earliest Pennsylvanian.

Meischner (1970) proposed a conodont zonation of the German Carboniferous.
is *Gnathodus tricarinatus deflectens* Zone is probably totally Pennsylvanian in
ze, and the lower part of that zone may correspond with the base of the *R.
rimus* Zone. The upper part of the *Gnathodus tricarinatus deflectens* Zone yielded
rms that are very close to *Idiognathoides sinuatus* and approximately corresponds
ith our *Neognathodus bassleri symmetricus* and *N. bassleri bassleri* Zones. The
se of his *Idiognathodus fiebigi* Zone may correspond with the base of our
iognathodus sinuosis Zone.

Palmieri (1969) reported the occurrence of Lower Pennsylvanian conodont faunas
om limestones in close proximity to sediments of the Wondai "Series," southeast
* Murgon, Queensland, Australia. The faunas are so poorly preserved that only
Lower or lower Middle Pennsylvanian age can be assigned to the collections.
owever, his faunas from samples 124A, 124, and 125 at outcrop B are at least
* young as our *Idiognathodus sinuosis* Zone.

Conodont Phylogeny

Lane (1967), Dunn (1970a, 1971), and Straka and Lane (1970) have discussed he phylogeny of Upper Mississippian and Lower Pennsylvanian conodonts. Dunn 1970a, p. 319) suggested that three major groups of Mississippian form-species ave rise to all the Pennsylvanian platform conodonts. These are the *Gnathodus irtyi* stock, the *Adetognathus unicornis* stock, and the *Gnathodus commutatus ommutatus* stock. Dunn (1971) later modified his views on the origin of *Idiognath-ides sinuatus* and suggested that the latter species derived from *Gnathodus defectus*. Ve do not agree with either of Dunn's postulations about the ancestry of *I. sinuatus* see Straka and Lane, 1970), but do agree in general with his three-stock plan s the origin of Pennsylvanian conodonts.

Although Dunn (1971) took issue with the paper by Straka and Lane (1970), ve still maintain that *Neognathodus bassleri symmetricus, Idiognathoides sinuatus,* nd *I. macer* appeared at approximately the same time (near the base of our *N. assleri symmetricus* Zone), and all were derived from the pre-existing and closely elated form-species *Idiognathoides noduliferus* and *I. sulcatus.*

The occurrences of the *Idiognathoides noduliferus* group—a concept that em-races *Idiognathoides noduliferus* (Ellison and Graves), *Idiognathoides japonicus* (go and Koike), and *Idiognathoides nevadensis* (Dunn)—is curious. We find that he group appeared in abundance at three different times in the Morrowan and wer Derryan. The first appearance was in lower Morrowan rocks as developed 1 Arkansas, Oklahoma, and Texas. This occurrence conforms with the *Idiognath-ides* aff. *I. noduliferus* Zone of Lane (1967) and the *Rhachistognathus primus* one (= *Spathognathodus muricatus* faunal unit) together with the *I. noduliferus* one of Lane and others (1971, 1972), Straka (1972), and herein. Higgins and ouckaert's (1968) occurrence of the form-species in the upper part of the *Homoceras eyrichianum* Zone (H1) through lower R2 strata in Belgium can be grouped in me with the forms from Arkansas, Oklahoma, and Texas. However, the material rom Belgium may be in part slightly older. In Nevada, *I. noduliferus* [= *I. nevadensis* Dunn)] occurs in abundance in the *N. bassleri bassleri* and *Idiognathodus sinuosis* ones. In Japan, *I. noduliferus* [= *I. japonicus* (Igo and Koike)] occurs with treptognathodus expansus* and thus is correlative with the middle Morrowan Nevada aunas. The second major occurrence of the species is therefore middle Morrowan 1 age as developed in Nevada and Japan. The third major occurrence of the pecies is in lower Derryan age rocks. The type specimen of *I. noduliferus* derives rom the Dimple Limestone in west Texas, and that formation is Derryan in age Sanderson and King, 1964). We have found Derryan occurrences of the form-species t the type Derryan section in New Mexico and in Derryan rocks in Nevada.

35

Whether these three time-separated occurrences of the form-species are artifact of an inadequate fossil record or represent three different origins of the form-specie are questions we cannot answer at this time. However, Koike's (1967) data woul suggest that the former is the proper interpretation. In any case, the type specime of the form-species *I. noduliferus* is from Derryan age rocks, and our usage c the name for lower and middle Morrowan occurrences must be considered wit some qualification.

Neognathodus bassleri appears in North America at the base of our *N. bassle symmetricus* Zone and ranges into the *Idiognathodus sinuosis* Zone. We have neve found the form-species in upper Morrowan rocks (*Idiognathodus klapperi* an *Idiognathoides convexus* Zones) in Arkansas, Oklahoma, Texas, New Mexico, c Nevada. However, *N. colombiensis* (Stibane), a form with close affinities to *N bassleri*, does occur in the lower Derryan of New Mexico and Nevada. Whethe one wishes to place *N. colombiensis* in synonymy with *N. bassleri* or not is totall a problem of nomenclature. However, none of the Derryan forms are consubspecifi with the Morrowan *N. bassleri symmetricus* or *N. bassleri bassleri*. Because c the apparent time separation between the occurrences of the two form-specie: a nonrelation must be considered. It is possible that *N. colombiensis* was derive from the Derryan occurrences of *I. noduliferus* in much the same manner as *N bassleri* was derived in the Morrowan.

Lane (1967) and Straka and Lane (1970) suggested that *Idiognathoides sinuatu* originated from *I. sulcatus* and *I. noduliferus*. We have never found transition: specimens in the correct stratigraphic position to suggest that *I. sinuatus* was derive from *Gnathodus commutatus commutatus* or *Gnathodus defectus* (Dunn, 1970: 1971).

The asymmetrically paired form-species (Class IIIb symmetry; Lane, 196: *Idiognathoides sinuatus* first appears slightly below the base of the *N. bassle symmetricus* Zone and ranges up to near the base of the *I. convexus* Zone. *A* that point, the right element of *I. sinuatus* gives rise to the right element of th asymmetrically paired (Class IIIb) form-species *I. convexus*. The full range c *I. convexus* is not known, but the asymmetrically paired (Class IIIb) form-specie *I. fossatus* (Branson and Mehl) occurs in Derryan age rocks and probably wa derived from *I. convexus*. The left element of all three form-species remains near the same throughout the Morrowan and into the Derryan.

The origins of the genera *Idiognathodus* and *Streptognathodus* are still problen atical. Dunn (1970a) and our collections from Nevada suggest that both were derive from the middle Morrowan occurrences of *Idiognathoides noduliferus*. Howeve the two genera are morphologically closest to *Idiognathoides sinuatus*, and a possib origin from that species cannot be discounted. Reported occurrences of *Idiognath dus* from the *Homoceras* Zone (Meischner, 1970, p. 1174) and *Streptognathod* from the *Homoceras* and *Reticuloceras* Zones (Higgins and Bouckaert, 1968, 18) in western Europe are homeomorphs of the respective later genera and a not part of the main idiognathodid and streptognathodid lineages.

We agree with Dunn's discussion of the evolution of the adetognathids an rhachistognathids and refer the interested reader to that work (Dunn, 1970a, 323–325). However, we find that *Rhachistognathus muricatus* ranges as high the *N. bassleri bassleri* Zone in Nevada.

Localities

Seventeen sections in northern Arkansas and northeastern Oklahoma and fifteen sections in southern Oklahoma of uppermost Mississippian (Chesterian) and Lower Pennsylvanian (Morrowan) have been sampled and processed for conodonts. The geographic location of each section is given in the following text, and all except secs. 14, 15, and 25 are illustrated on Figures 3-5. Graphic measured sections (Figs. 11-23) are provided for secs. 1-13, and lithologic descriptions of each section are on the microfiche appendix.[1] Graphic measured sections of Localities 17 and 18 and 20-25 are on Figure 6, and information on sample locations is given in the microfiche appendix. Secs. M-5, M-25, M-26, M-28, M-42, M-51A, M-51, and M-64 are illustrated on Figure 7 and described by Sutherland and Henry (1974). Information concerning the sample locations for these localities is also given in the microfiche appendix. Localities 14-16 and 19 represent spot samples or small stratigraphic intervals, and only the geographic information and, where possible, footage location within each section are provided. Samples listed on the microfiche card but not included on Figures 24-31 were barren of conodonts.

LOCALITY 1

E1/2NW1/4NW1/4 sec. 25, T. 3 S., R. 1 E., Springer quadrangle, Carter County, Oklahoma. Locality 60 of Tomlinson (1959, p. 320); "Caddo Village" locality of Elias (1956a, p. 98). This locality is south of the north turn of the road 0.8 mi (1.3 km) west-northwest of the abandoned Caddo Village. Sandstone and shale units of the Primrose Member of the Golf Course Formation are exposed in an intermittent stream valley that cuts across the strike of the beds and flows to the southwest. The section begins at the highest exposed sandstone unit of the Primrose and continues downsection.

---

[1]Microfiche appendix is in pocket inside back cover. Order additional photocopies or microfiche from ASIS/NAPS, c/o Microfiche Publications, 305 E. 46th Street, New York, N.Y. 10017. Refer to NAPS document 02291.

LOCALITY 2

S1/2NW1/4SW1/4 and S1/2SE1/4NE1/4SE1/4 sec. 4, T. 3 S., R. 2 E., Gene Autry quadrangle, Carter County, Oklahoma. Locality 61 of Tomlinson (1959, p. 321); type locality of the Lake Ardmore Formation of Tomlinson and McBee (1959), p. 11). Four sandstone ridges exposed east of a private entrance road represent the three sandstone units of the Lake Ardmore Sandstone, and the fourth, southernmost, is the Primrose Sandstone. Topographic low areas covered by dense vegetation between the ridges probably represent shaly units or less resistant sandstones. The private entrance road is 2 mi (3.2 km) east of Highway 77 and the town of Springer. The section begins at the highest exposure of the Primrose and continues downsection to the second Lake Ardmore Sandstone unit.

LOCALITY 3

E1/2NW1/4SE1/4 sec. 1, T. 3 S., R. 2 E., Gene Autry quadrangle, Carter County, Oklahoma. Localities 60 and 63 of Tomlinson (1959, p. 320-322); type locality of Gene Autry Shale Member of Golf Course Formation of Elias (1956a, p. 99); the "cephalopod" locality of Goldston (1922, p. 7), Girty and Roundy (1923, p. 334), and Tomlinson (1929, p. 19). A continuous exposure of strata occurs in a north-south tributary stream to Caddo Creek. The tributary stream parallels on the west side the Gulf, Colorado, and Santa Fe Railroad embankment. These strata include, in ascending order, sandy shale, which is interpreted to be the lateral facies of the Lake Ardmore Sandstone Member and the Academy Church Shale Member of the Springer Formation, and the Primrose and Gene Autry Members of the Golf Course Formation. The locality is about 5.5 mi (8.8 km) east of Springer and approximately 2 mi (3.2 km) north of Gene Autry, Oklahoma. The section begins at the highest exposed Primrose and continues downsection to the lowest exposure of the lateral shale facies of the Lake Ardmore.

LOCALITY 4

N1/2NW1/4NE1/4SW1/4 sec. 4, T. 3 S., R. 2 E., Gene Autry quadrangle, Carter County, Oklahoma (new locality); 75 ft (22.9 m) of shale ("B" Shale) are exposed in a small gully developed 65 ft (19.8 m) south of "Overbrook" Sandstone ridge. The section begins at the lowest exposure of the "B" Shale. The exposure can be reached by following the private entrance road from Locality 2, through the closed gate, and continuing about 600 ft (182.9 m) north of the farmhouse.

LOCALITY 5

Center of the NE1/4SE1/4 and center of the W1/2SW1/4SE1/4NE1/4 sec 3, T. 3 S., R. 2 E., Gene Autry quadrangle, Carter County, Oklahoma. Type locality of Target Limestone of Bennison (1954, p. 913); Stop No. 14, Ardmore

Geological Society field trip (1966). Target embankment of the Ardmore Air Force Base is visible at the foot of the mountains. "Overbrook" Sandstone exposed under and on either side of a small, collapsed, wooden bridge in the road to the targets, about 500 ft (152.4 m) north of the Target Limestone exposure. Limestone exposed here is 3 to 5 ft (0.91 to 1.52 m) thick, but topographic relief to either side of the outcrop suggests presence of 15 to 20 ft (4.6 to 6.1 m) total thickness. The Primrose Sandstone crops out 450 to 500 ft (137.2 to 152.4 m) south of the Target Limestone. The section begins at the top of the Target and continues down to the base of the "Overbrook."

LOCALITY 6

W1/2NW1/4SE1/4 sec. 4, T. 3 S., R. 2 E., Gene Autry quadrangle, Carter County, Oklahoma. Locality 64 of Tomlinson (1959, p. 322). Ridge of "Overbrook" Sandstone exposed parallel to road. Prominent vertical standing cliff of lowermost Lake Ardmore Sandstone can be seen about 0.25 mi (0.4 km) to the south of this outcrop. The section begins at the base of the "Overbrook" Sandstone and continues up to the second sandstone unit of the Lake Ardmore Member.

LOCALITY 7

Center of NW1/4NE1/4NE1/4 sec. 2, T. 3 S., R. 1 E., Springer quadrangle, Carter County, Oklahoma. Locality 65 of Tomlinson (1959, p. 322). We interpret the 34 ft (10.4 m) of strata exposed in a west cutbank of Tulip Creek, just south of Highway 53 West, to be not a shale above the Rod Club as Tomlinson (1959, p. 322) suggested, but a second exposure of the Tiff Member of the Goddard Formation. The section begins at stream level and continues upward to the top of the exposure.

LOCALITY 8

NE1/4SE1/4SW1/4 and SW1/4NW1/4SE1/4 sec. 16, T. 3 S., R. 1 E., Springer quadrangle, Carter County, Oklahoma. Locality 66 of Tomlinson (1959, p. 323). Vertical standing beds of Rod Club and "Overbrook" Sandstone Members of the Springer Formation are exposed, and the section begins at the base of the Rod Club and continues to the highest exposure of "Overbrook."

LOCALITY 9

E1/2NE1/4SE1/4NE1/4 and E1/2SE1/4NE1/4NE1/4 sec. 12, T. 4 S., R. 1 E., Ardmore West quadrangle, Carter County, Oklahoma. Locality 66 of Tomlinson (1959, p. 323). Rod Club Sandstone Member of the Springer Formation and uppermost Goddard Shale exposed on both sides of Highway 77, about 0.33 mi (0.53 km) north of the Ardmore City Filtration Plant Road. The section begins at the lowest exposure of the Goddard and continues to the highest exposure of the Rod Club.

LOCALITY 10

SE1/4NW1/4SE1/4SE1/4 and NE1/4SW1/4SE1/4SE1/4 sec. 1, T. 4 S., R. 1 E., Ardmore West quadrangle, Carter County, Oklahoma. Locality 67 of Tomlinson (1959, p. 323). Upper shales of Goddard Formation exposed in cutbank of creek (now emponded). Located about 0.5 mi (0.8 km) below City Lake Dam on the Noble Ranch, 0.2 mi (0.32 km) west of Highway 77. Section begins at the base of the cutbank and continues to the top.

LOCALITY 11

E1/2NE1/4SE1/4SE1/4 and E1/2SE1/4SE1/4SE1/4 sec. 16; and NW1/4SW1/4SW1/4SW1/4 sec. 15, T. 3 S., R. 1 E., Springer quadrangle, Carter County, Oklahoma. Locality 68 of Tomlinson (1959, p. 323) and type locality of Tiff Member (=Grindstone Creek Member) of Goddard Formation of Tomlinson (1959, p. 323). There are extensive exposures of the middle Goddard Shale (="D" Shale) above and below the Tiff unit, in an east cutbank of Grindstone Creek. A lake created by a newly constructed dam on Grindstone Creek has effectively flooded the entire exposure of the Tiff strata, and only the uppermost bluffs of the high cutbank bear shale exposures which may be accessible by boat. The section begins at the lowest and continues to the highest exposures of the Goddard.

LOCALITY 12

NW1/4SW1/4SW1/4 sec. 19, T. 2 S., R. 1 W., Carter County, Oklahoma. Locality 71 of Tomlinson (1959, p. 324); type locality of Redoak Hollow Member of Goddard Formation of Elias (1956a, p. 85); 66 ft (20.1 m) of highly iron-stained, very thinly bedded sandstone alternating with shale beds are exposed in creek bed and across the pasture road. Elias' collecting site "IV" was sampled and is found by turning left at the first fork in the pasture road upon emerging from the woods. Outcrop material can be seen in road bed exactly 0.3 mi (0.5 km) along the left fork from the fork in the road. Section begins at lowest and continues to highest exposures.

LOCALITY 13

N1/2SW1/4SW1/4SE1/4 sec. 25, T. 2 S., R. 1 E., Springer quadrangle, Carter County, Oklahoma. Locality 72 of Tomlinson (1959, p. 324). The Goddard-Caney contact occurs in a north cutbank of Tulip Creek, approximately 100 yds (91.4 m) north of the first blacktop road to the right upon emerging from the Arbuckle

Mountains on Highway 77. The contact here was arbitrarily placed at the top of a 12 in. (0.3 m) limestone bed (samples 13–17) in the midst of a shale succession that strikes parallel to the mountain front. Shale above the limestone unit is considered basal Goddard, although hard, blocky Caney-type lithology is interbedded with characteristic soft Goddard Shale. The section begins at the limestone in the top of the Caney and continues up to the highest exposed Goddard.

LOCALITY 14

NE1/4NE1/4NE1/4 sec. 30, T. 3 N., R. 7 E., Stonewall quadrangle, Pontotoc County, Oklahoma. The shale (Rhoda Creek?) below the base of the Union Valley Sandstone, and the Union Valley Sandstone are exposed in an east cutbank of Rhoda Creek, about 70 yd (64 m) south of the first paved road running west of Highway 3, approximately 1 mi (1.6 km) northwest of the town of Union Valley. All 10 ft (3.05 m) of exposed shale below the Union Valley were sampled.

LOCALITY 15

NE1/4NE1/4NE1/4 sec. 29, T. 3 N., R. 7 E., Pontotoc County, Oklahoma. This locality was visited but not sampled. Matrix from several specimens of *Gastrioceras* in the University of Iowa collections from this locality was processed.

LOCALITY 16

Uppermost Pitkin Limestone and lowermost Hale Formation were collected on the southeast side of State Highway 10 as it ascends the northern slope of Braggs Mountain, NW1/4SE1/4NE1/4 sec. 29, T. 15 N., R. 20 E., Muskogee County, Oklahoma.

LOCALITY 17

Uppermost Pitkin Limestone and lowermost Prairie Grove Member of the Hale Formation were collected in a bluff on the east side of State Highway 59, north of the town of Davidson, SW1/4NW1/4SW1/4 sec. 23, T. 12 N., R. 33 W., Crawford County, Arkansas.

LOCALITY 18

Samples were collected from the Prairie Grove Member of the Hale Formation in the bed of a small intermittent stream SE corner sec. 1, T. 14 N., R. 31 W., about 100 yd (91.44 m) south of the West Fork–Moffit Road and about 2.5 mi (4 km) southwest of the Highway 71 junction at West Fork, Washington County,

Arkansas. This is the type locality of *Reticuloceras wainwrighti* Quinn (1966, p. 14). Sample 3 was collected from the goniatite horizon in the stream bed about 30 yd (27.4 m) upstream from samples 1 and 2. Samples 1 and 2 are measured in feet above the base of the section.

LOCALITY 19

This is locality L117 of Quinn and Saunders (1968, p. 400). Matrix from ammonoids collected by W. B. Saunders from the conglomerate at the top of a siltstone and shale section that rests on the Fayetteville Shale was processed. Saunders (1969, oral commun.) considers this conglomerate to be part of the Cane Hill Member of the Hale Formation.

LOCALITY 20

Exposures on the east side of Fort Gibson Dam along the east side of State Highway 16, SW1/4NE1/4 sec. 18, T. 16 N., R. 20 E., Cherokee County, Oklahoma. The section starts at the east end of the dam and extends up the bluff. Fayetteville through Atoka strata are exposed, and 23 ft (7 m) of the Bloyd and 41 ft (12.5 m) of the Hale are continuously exposed and were sampled. Also, the upper 4 ft (1.2 m) of the Pitkin were collected.

LOCALITY 21

Bluff on the east side of Fort Gibson Reservoir about 0.25 mi (0.4 km) north of the east end of State Highway 51 bridge, E1/2SW1/4SE1/4 and W1/2NE1/4SE1/4 sec. 23, T. 17 N., R. 19 E., Cherokee County, Oklahoma. The upper 2 ft (0.61 m) of the Hale and 38 ft (11.6 m) of the Bloyd are exposed and were sampled. The bluff is capped by Atokan rocks.

LOCALITY 22

Kansas, Oklahoma, and Gulf Railroad cut in S1/2S1/2 sec. 17, T. 19 N., R. 19 E., Mayes County, Oklahoma. The upper 6 ft (1.8 m) of the Hale and 19 ft (5.8 m) of the Bloyd are exposed and were sampled.

LOCALITY 23

Locality and sample information for the Evansville Mountain exposure are the same herein as those for Lane (1967, p. 925-926); however, additional sample intervals are included in this report.

LOCALITY 24

Locality and sample information for the type Brentwood section herein are the same as those in Lane (1967, p. 926). However, additional sample intervals are included in this report.

LOCALITY 25

Locality information and sample intervals for the Peyton Creek section herein are the same as those of Lane (1967, p. 925; T. 3 N. in this reference should be T. 13 N.). Three additional sample intervals (2, 4, 5) are included in this report.

LOCALITY M-5

Exposures in the Betsy Lee Creek bed and Jeffries Quarrysite, NW1/4NW1/4 sec. 33, T. 13 N., R. 20 E., Muskogee County, Oklahoma. P. K. Sutherland and Thomas W. Henry (1972, oral commun.) informed us that the Jeffries Quarrysite locality is now obliterated. In the stream bed, the sampled section starts at the top of the Pitkin Limestone and continues through the Hale into the lower part of the Bloyd. One sample (13) was collected from upper Bloyd in the now-obliterated Jeffries Quarrysite.

LOCALITY M-25

East side of State Highway 10 as it descends Parkhill Mountain toward Tahlequah, NE1/4NE1/4SE1/4 sec. 28, T. 16 N., R. 22 E., Cherokee County, Oklahoma. The upper few feet of the Pitkin and lower several feet of the Hale were collected.

LOCALITY M-26

Exposures along the south side of State Highway 82, northeast of Cookson, SW1/4 sec. 6, T. 14 N., R. 23 E., and SE1/4 sec. 1, T. 14 N., R. 22 E., Cherokee County, Oklahoma, were sampled. Sample 1 was collected from the lowest exposures in the ditch on the south side of the road near the bottom of the hill. The Hale-Bloyd lithologic contact at this locality is difficult to pick, and its placement herein represents only a suggestion based on faunal correlations to the type section.

LOCALITY M-28

Bluff on the west side of the Arkansas River overlooking the Webbers Falls Lock and Dam Project. The section begins in the center of the NE1/4 sec. 33,

T. 13 N., R. 20 E., extends into NE1/4SW1/4NE1/4 sec. 33, T. 13 N., R. 20 E., and terminates in NW1/4NE1/4 sec. 33, T. 13 N., R. 20 E., Muskogee County, Oklahoma. The exposure starts in the Pitkin Limestone and extends through the Morrowan. Atokan strata are exposed near the top of the bluff. The Hale-Bloyd contact at this locality is difficult to pick, and its placement represents only a suggestion based on faunal correlations to the type sections.

LOCALITY M-42

North bank of Greenleaf Creek and up the bluff on the north side of the dirt road that leads out onto the dam, SW1/4NE1/4 sec. 10, T. 13 N., R. 20 E., Muskogee County, Oklahoma. The base of the section starts within the lower Bloyd, and Atokan rocks cap the bluff. Sample 5 is offset east from sample 4 and represents the same horizon. A major 260 ft (79.2 m) eastward offset in the section occurs between samples 9 and 10.

LOCALITIES M-51A AND M-51

West bank of the Illinois River at the northern end of Tenkiller Ferry Reservoir, on and around Singin' T Acres, property of the country and western singer, Hank Thompson. Samples 1-4 were collected in the SE1/4NE1/4 sec. 7, T. 15 N., R. 23 E., about 0.25 mi (0.4 km) north of the main part of the section and represents section M-51A. The base of the main part of the section (M-51) starts at normal river level, SE1/4NE1/4SE1/4 sec. 7, T. 15 N., R. 23 E., about 100 yd (91.44 m) north of the boathouse, and extends up the bluff to the parking lot and garden of Singin' T Acres. The section is then offset southward and extends up the shale and limestone slope just west of the garden, NW1/4SE1/4SE1/4 sec. 7, T. 15 N., R. 23 E., to the base of the Atoka. All above locations are in Cherokee County, Oklahoma. The Hale-Bloyd contact at this locality is difficult to pick, and its placement herein represents only a suggestion based on faunal correlations with the type sections.

LOCALITY M-64

The Northwest Qualls section starts in the intermittent stream, NW1/4NE1/4SE1/4 sec. 34, T. 15 N., R. 21 E., and extends up the bluff on the north side of the creek to the Morrowan-Atokan contact, NE1/4NE1/4SE1/4 sec. 34, T. 15 N., R. 21 E., Cherokee County, Oklahoma.

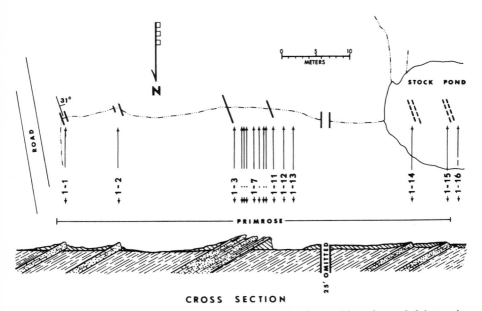

CROSS SECTION

Figure 11. Diagrammatic illustration of the outcrop showing position of sampled intervals at Locality 1, Carter County, Oklahoma.

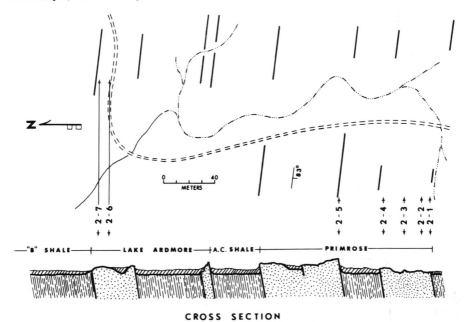

CROSS SECTION

Figure 12. Diagrammatic illustration of the outcrop showing position of sampled intervals at Locality 2, Carter County, Oklahoma.

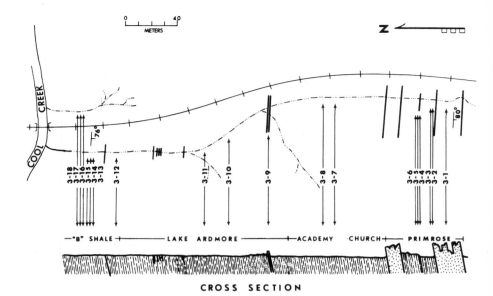

Figure 13. Diagrammatic illustration of the outcrop showing position of sampled intervals at Locality 3, Carter County, Oklahoma.

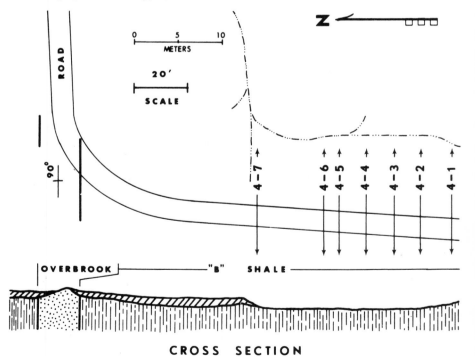

Figure 14. Diagrammatic illustration of the outcrop showing position of sampled intervals at Locality 4, Carter County, Oklahoma.

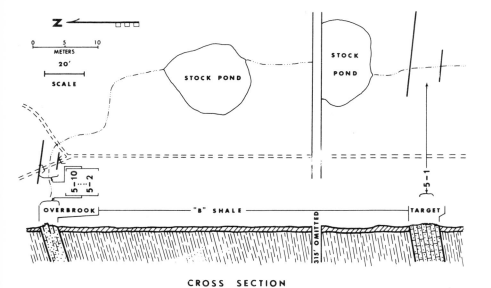

CROSS SECTION

Figure 15. Diagrammatic illustration of the outcrop showing position of sampled intervals at Locality 5, Carter County, Oklahoma.

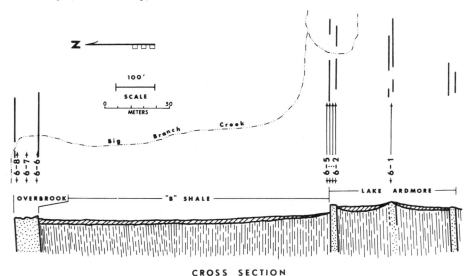

CROSS SECTION

Figure 16. Diagrammatic illustration of the outcrop showing position of sampled intervals at Locality 6, Carter County, Oklahoma.

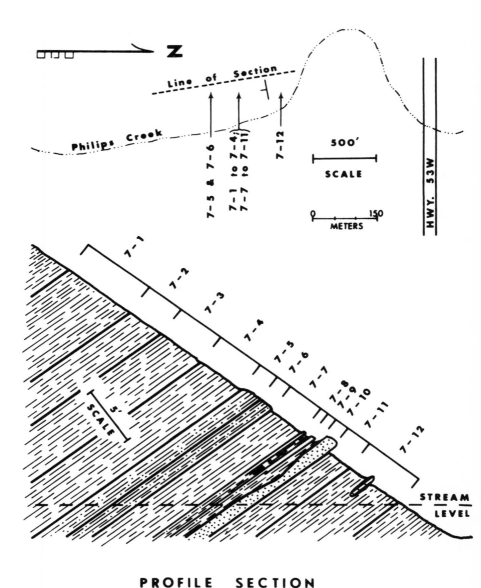

PROFILE SECTION

Figure 17. Diagrammatic illustration of the outcrop showing position of sampled intervals at Locality 7, Carter County, Oklahoma.

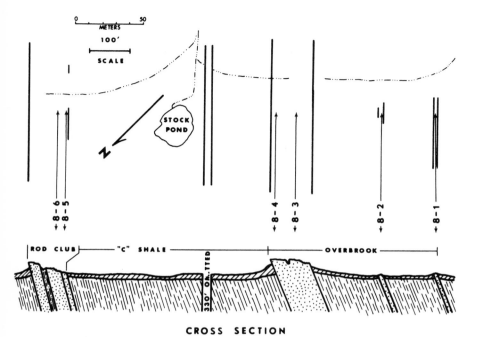

Figure 18. Diagrammatic illustration of the outcrop showing position of sampled intervals at Locality 8, Carter County, Oklahoma.

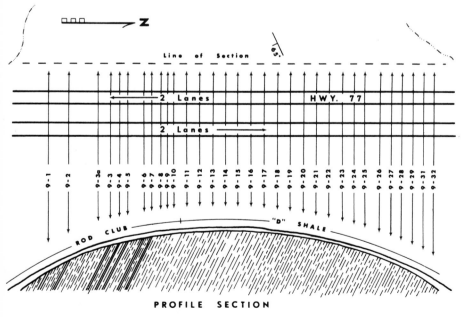

Figure 19. Diagrammatic illustration of the outcrop showing position of sampled intervals at Locality 9, Carter County, Oklahoma.

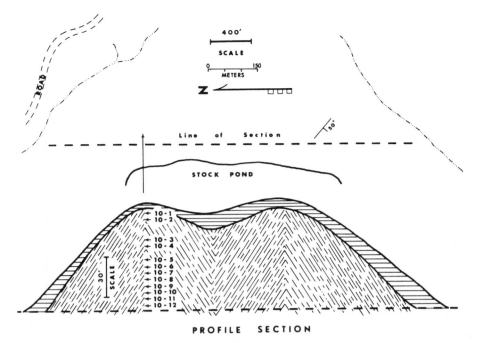

Figure 20. Diagrammatic illustration of the outcrop showing position of sampled intervals at Locality 10, Carter County, Oklahoma.

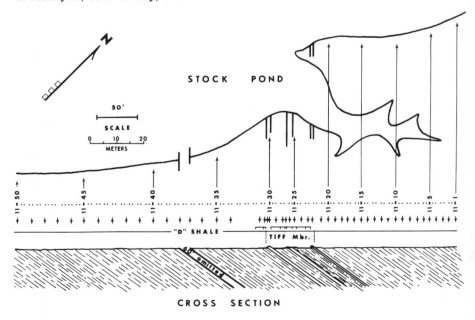

Figure 21. Diagrammatic illustration of the outcrop showing position of sampled intervals at Locality 11, Carter County, Oklahoma.

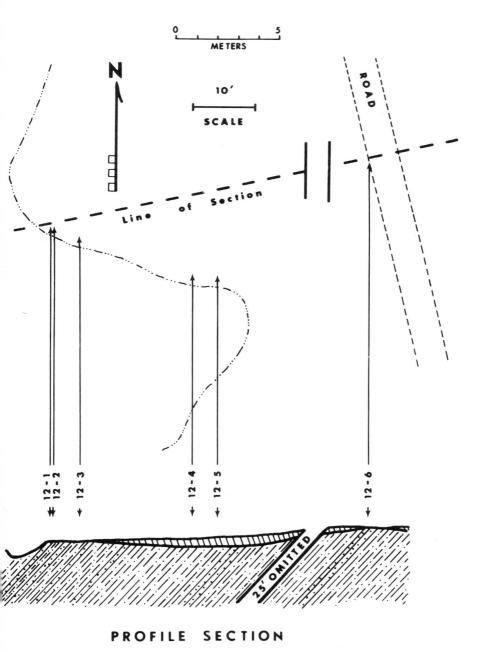

PROFILE SECTION

Figure 22. Diagrammatic illustration of the outcrop showing position of sampled intervals at Locality 12, Carter County, Oklahoma.

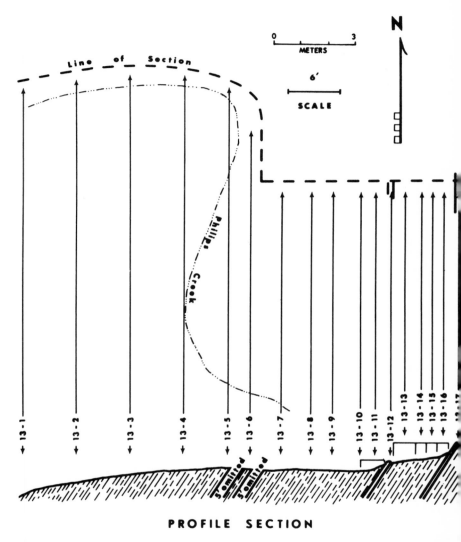

Figure 23. Diagrammatic illustration of the outcrop showing position of sampled interval at Locality 13, Carter County, Oklahoma.

Systematic Paleontology

All available platform form-species are treated herein. The bars and blades seem to be of less stratigraphic value and are not given full consideration in this study. This study was completed before multielement taxonomy gained popularity, and consequently little attempt is made in establishing multielement associations. However, the form nomenclature presented here will have a direct affect on apparatus nomenclature when multielement associations are established.

REPOSITORIES

Specimens figured are reposited in the collections of the University of Iowa (SUI), the United States National Museum (USNM), the University of Missouri (UM and MSM), and the Bureau of Economic Geology at the University of Texas (BEG).

CLARIFICATION OF SPECIES CONCEPT

Previous concepts of conodont form-species have been based on individual elements with a single holotype or lectotype established as reference for the form-species. Lane (1968) presented a symmetry classification of conodont element-pairs and stated that each conodont species displays a characteristic element-pair symmetry type. Furthermore, he recognized that two form-species (one exclusively left-sided, and one exclusively right-sided) may belong to one asymmetrical conodont element-pair (Class IIIb). In the present study, it is believed that maintenance of two form-species belonging to one element-pair of the conodont-bearing organism only burdens the literature, clouds the true relations, and offers no advantage for biostratigraphic refinement. Therefore, in this study, the two asymmetrical form-species of a pair are synonymized. In effect, this places the specific concept and, necessarily, the generic concept on an element-pair basis, rather than on the conventional form-species, single-element basis, and more closely approximates the correct multielement association. In such a system, conodont form-species nomenclature is affected only where both lefts and rights of a Class IIIb symmetry pair have been given different names.

The symmetry class designations for element-pairs used herein are after Lane (1968). Left and right refer to curvature and are determined by orienting a specimen in upper view with the anterior end at the top. A left specimen is one in which the left edge is the outer side and the blade-carina is convex toward that side. A right specimen is one in which the right edge is the outer side and the blade-carina is convex toward that side. Figures 24–31 illustrate sample numbers and numerical counts of lefts (upper) and rights (lower) of all conodont form-taxa studied for this investigation.

Our usage of left-sided and right-sided differs fundamentally from the definition of left and right given above. A left-sided specimen is one in which a particular morphologic development (that is, process, lobe, and so on) is invariably on the specimen's left side, regardless of its direction of curvature. Determination of the "sidedness" of a specimen must be done with it oriented in the manner discussed earlier. Thus, the form-species *Rhachistognathus muricatus* (Dunn; Class IIIa) contains both rights and lefts, but the form-species is strictly left-sided.

The locality and sample horizon of each figured specimen are given after the repository number in the plate descriptions. The first number refers to the location as given in the Appendix, and the second number is the sample from which the specimen derives in that section.

TAXON																									
FORMATION	SPRINGER																								
MEMBER	"B" SHALE							TARGET	LAKE ARDMORE									ACADEMY CHURCH							
LOC.	4							5	3																
				B									B				B						B		B
SAMPLE	1	2	3	4	5	6	7	1	17	16	15	14	13	12	11	10	9	8	7	6	5	4	3	2	1
Gnathodus girtyi intermedius	0/1	0/1	0/2		0/3	0/2	1/2																		
Rhachistognathus muricatus								71/46																0/1	
Rhachistognathus primus								11/18																	
Adetognathus lautus								17/15	2/2	1/0	1/1	1/3		0/2	0/1										
Idiognathoides noduliferus								1/0								1/0					1/0	36/40		4/3	
I. noduliferus M.T.S. B										0/1												0/31		0/4	
I. noduliferus M.T.S. C																						24/0		8/0	
I. noduliferus M.T.S. A																						2/1			
I. noduliferus M.T.S. D																						6/9		1/0	
I. noduliferus M.T.S. E								1/0													0/1	5/5		1/1	
Idiognathoides sulcatus subsp.																		1/1	0/1	0/2		11/9		7/10	
I. sulcatus subsp. M.T.S. F																				0/2		0/10		0/11	
I. sulcatus subsp. M.T.S. G																									
I. sulcatus subsp. M.T.S. A																		1/1				1/0			
I. sulcatus subsp. M.T.S. H																						1/1		0/1	
I. sulcatus subsp. M.T.S. I																					0/1			1/3	
Idiognathoides macer																						0/1			
Idiognathoides sinuatus																						0/1		0/1	
Neognathodus bassleri symmetricus																									
Neognathodus bassleri bassleri																									
Neognathodus sp.																									
Idiognathodus sinuosis																									
Streptognathodus sp.																									

TAXON																		
FORMATION	GOLF COURSE																	
MEMBER	PRIMROSE																	
LOC.	1															2		
				B														B
SAMPLE	16	15	14	12/13	11	10	9	8	7	6	5	4	3	2	1	3	2	1
Gnathodus girtyi intermedius																		
Rhachistognathus muricatus																		
Rhachistognathus primus																		
Adetognathus lautus	43/25	1/4				1/0											1/1	
Idiognathoides noduliferus	55/42	5/0	1/0						2/1	1/0			0/1					
I. noduliferus M.T.S. B	0/25	0/3	0/1															
I. noduliferus M.T.S. C	17/0	5/0																
I. noduliferus M.T.S. A	8/0																	
I. noduliferus M.T.S. D	12/4																	
I. noduliferus M.T.S. E	5/7	4/2																
Idiognathoides sulcatus subsp.	392/381	11/6			1/0	12/4		1/0	2/2	3/2			1/0	1/0				
I. sulcatus subsp. M.T.S. F	0/74	0/8																
I. sulcatus subsp. M.T.S. G		8/0			2/0													
I. sulcatus subsp. M.T.S. A	8/2																	
I. sulcatus subsp. M.T.S. H	55/46	3/0																
I. sulcatus subsp. M.T.S. I	15/41	3/11																
Idiognathoides macer	1/2					1/0			1/1									
Idiognathoides sinuatus	13/31	38/38	2/1		1/0	13/40	1/4	2/11	284/366	24/28	11/10	2/0	3/3		0/3		0/3	
Neognathodus bassleri symmetricus		68/58																
Neognathodus bassleri bassleri					1/0	10/16	1/5	9/6	156/200	6/7	4/8	3/4	2/5	1/0	1/2			
Neognathodus sp.		1/0								1/0								
Idiognathodus sinuosis															3/0	0/1	36/48	
Streptognathodus sp.																	0/1	

Figure 24. Distribution of conodont taxa in sampled Localities 1, 2, 3, 4, and 5. $\frac{0}{0}$ = number of left specimens above number of right specimens. B above sample numbers means the samples were barren of conodonts.

FORMATION	GODDARD											
MEMBER				TIFF								
LOC.	7											
TAXON / SAMPLE	B 12	11	10	9	8	7	6	5	4	3	2	1
Gnathodus commutatus subsp		4/7	2/2	11/11	2/2	4/7	5/7	7/8	33/43	5/4	24/35	1/0
Neoprioniodus singularis			1/0	1/1							16/9	0/3
Geniculatus claviger				2					2		2	
Cavusgnathus unicornis							0/1		0/3			
Gnathodus bilineatus morph. ?		8/7		10/11	5/16	10/16	14/17	20/19	10/7	3/0	45/38	0/1
G. bilineatus morphotype α		1/0	1/0	5/4	5/0	1/0	6/4	7/6	7/9	5/3	16/11	1/2
morphotype β					1/2	2/4	2/2	2/1	8/2		4/9	
morphotype γ		2/2		3/1		6/5	3/1	1/2	7/5	1/1	4/2	
morphotype δ		0/1		1/0								
Cavusgnathus naviculus												0/1
Ozarkodina roundyi		1/0					1/1				0/1	
Neoprioniodus scitulus				0/1	0/1		1/2		7/5		3/4	0/1
Adetognathus unicornis											0/1	
Ozarkodina compressa												
Neoprioniodus sp.												
Cavusgnathus altus s.l.												
Neoprioniodus varians												
Neoprioniodus camurus												
Hibbardella milleri												
Spathognathodus minutus												
Ozarkodina delicatula												
Rhachistognathus primus												
Adetognathus lautus												

FORMATION	GODDARD																	
LOC.	10						9											
TAXON / SAMPLE	F 8/12	7	B 5/6	4	B 3	F 1/2	B 27/31	26	25	F 20/24	19	18	17	B 16	15	F 13/14	12	B 11
Gnathodus commutatus subsp		0/1		3/2														
Neoprioniodus singularis																		
Geniculatus claviger																		
Cavusgnathus unicornis																		
Gnathodus bilineatus morph. ?											0/1		3/1		1/0			
G. bilineatus morphotype α												1/0			1/0			
morphotype β																		
morphotype γ																		
morphotype δ																		
Cavusgnathus naviculus								0/1	0/1									
Ozarkodina roundyi									0/1		0/1		1/0					
Neoprioniodus scitulus				1/0							0/1							
Adetognathus unicornis											1/0						0/1	
Ozarkodina compressa																		
Neoprioniodus sp.																		
Cavusgnathus altus s.l.																		
Neoprioniodus varians																		
Neoprioniodus camurus																		
Hibbardella milleri																		
Spathognathodus minutus																		
Ozarkodina delicatula																		
Rhachistognathus primus																		
Adetognathus lautus																		

FORMATION	SPR.			PITKIN								HALE	
MEMBER	R.C.											P.G.	
LOC.	9			25								17	
TAXON / SAMPLE	B 9/10	8	B 1/7	1	2	3	6	7	8	9	10	1	2
Gnathodus commutatus subsp		1/0		0/1					0/1				2/0
Neoprioniodus singularis													
Geniculatus claviger													
Cavusgnathus unicornis						0/1	0/1	0/2	0/2				
Gnathodus bilineatus morph. ?													
G. bilineatus morphotype α		1/0											
morphotype β													1/0
morphotype γ													
morphotype δ				1/0									
Cavusgnathus naviculus				0/3		0/1	0/1		0/3	0/1	0/2		
Ozarkodina roundyi		0/1											
Neoprioniodus scitulus									2/4		2/0		
Adetognathus unicornis		1/0		6/0	1/3	0/2	5/1	1/0	9/14	2/3	5/5		
Ozarkodina compressa				1									
Neoprioniodus sp.				0/1	0/2								
Cavusgnathus altus s.l.							0/1	0/1	0/2		0/1	0/2	
Neoprioniodus varians						0/1							
Neoprioniodus camurus							1/0	1/1	0/2	1/0	1/3		
Hibbardella milleri									1	1	1		
Spathognathodus minutus									1				
Ozarkodina delicatula									1				
Rhachistognathus primus													1/0
Adetognathus lautus													2/6

Figure 25. Distribution of conodont taxa in sampled Localities 7, 9, 10, 17, and 25. $\frac{0}{0}$ **= number of left specimens above number of right**

FORMATION	GODDARD																																										
MEMBER																																	TIFF										
LOC.	13																11																										
						B		B	F								B			F		F		F		F						B						F			B		F
TAXON / SAMPLE	17	16	15	14	13	12	11	9/10	8	7	6	5	4	3	2	1	49/50	48	47	46	45	41/44	40	39	38	34/37	33	32	31	30	29	27/28	26	25	24	23	22	15/21	14	13	6/12	5	1/4
Gnathodus commutatus subsp.	2/0		1/0		0/2		1/3			0/1	1/1	2/3	7/6	1/1	1/1												58/78	41/36	4/5	5/3				0/1	2/2	2/2	4/3		1/0	1/0			
Neoprioniodus singularis		1/0					1/0				0/1		3/4														1/0												0/1				
Geniculatus claviger				1			1				1																	4								1							
Cavusgnathus unicornis		0/1								0/3																																	
Gnathodus bilineatus morpho. ?	0/1											1/0											0/1		1/0		39/45	59/60	9/12		1/0		2/2	3/3		2/5	3/3					0/1	
G. bilineatus morphotype α					1/0		0/1											0/1									26/27	23/25	3/5	3/0			0/1	0/2	6/21	4/3						0/1	
morphotype β																		0/1									40/17	10/17	1/0						1/0								
morphotype γ																					1/0						23/11	9/13	2/2	1/0			1/0	1/0	4/2	4/3	0/1						
morphotype δ																											0/3		0/3					1/1		4/2							
Cavusgnathus naviculus													0/1			0/1																											
Ozarkodina roundyi													0/1		1/0													1/1															
Neoprioniodus scitulus																			0/1				1/0					4/3									1/3		1/0				
Adetognathus unicornis																											1/0																

Figure 26. Distribution of conodont taxa in sampled Localities 11 and 13. $\frac{0}{0}$ = number of left specimens above number of right specimens. B and F above sample numbers mean the samples were barren of conodonts, or fragmentary remains were recovered, respectively.

Group	Taxon	Loc. 16, sample 1	Loc. 16, sample 2	Loc. M-25, sample 1	Loc. M-25, sample 2	Loc. M-25, sample 3	Loc. 18, sample 1	Loc. 18, sample 2	Loc. 18, sample 3
	FORMATION	PIT.	HALE	PIT.	HALE		HALE		
	MEMBER						PRAIRIE GR.		
	Cavusgnathus unicornis	0/4			0/1				
	Cavusgnathus altus s.l.	0/12							
	Gnathodus commutatus subsp.	0/1							
	Rhachistognathus primus		2/2		1/0			0/1	
	Adetognathus lautus		3/3		7/5	2/2	2/0	4/8	
I. noduliferus	*Idiognathoides noduliferus*					1/0	0/1	0/1	
I. noduliferus	M.T.S. B		0/5		0/3		0/1	0/4	
I. noduliferus	C		4/0		2/0			1/0	
I. noduliferus	A		3/1						
I. noduliferus	D		2/1						
I. noduliferus	E		4/2		2/1	1/0		2/0	1/0
I. sulcatus subsp.	*Idiognathoides sulcatus* subsp.				1/3			1/0	
I. sulcatus subsp.	M.T.S. F		0/1		0/2	0/1			
I. sulcatus subsp.	G				5/0	1/0		1/0	
I. sulcatus subsp.	A		2/0		1/0			2/0	
I. sulcatus subsp.	H				3/3	1/0			
I. sulcatus subsp.	I				3/7	1/3			
	Ozarkodina delicatula		1						
	Gnathodus bilineatus morpho. γ			2/0					
	Idiognathoides sinuatus					5/6			
	Idiognathoides sulcatus parvus								
	Adetognathus spathus				2/1	1/2	0/1	2/2	
	Neognathodus bassleri symmetricus								
	N. bassleri bassleri								
	Idiognathodus sinuosis								
	Spathognathodus minutus								
	Spathognathodus n. sp.								
	Streptognathodus expansus								

Group	Taxon	24-0	24-3	24-5	24-6	24-9	24-10	24-11	24-13	24-15	24-18	24-19	24-20	24-21	24-22	24-23	24-24	24-25
	FORMATION	HALE					BLOYD											
	MEMBER	PRAIRIE GR.					BRENTWOOD									WOOLSEY		
	Cavusgnathus unicornis																	
	Cavusgnathus altus s.l.																	
	Gnathodus commutatus subsp.																	
	Rhachistognathus primus																	
	Adetognathus lautus	3/3	1/1	3/0	1/1	0/1	0/1		1/5	1/1	2/0	18/23	28/21	18/13	17/15	25/12	12/13	1/0
I. noduliferus	*Idiognathoides noduliferus*																	
I. noduliferus	M.T.S. B																	
I. noduliferus	C																	
I. noduliferus	A																	
I. noduliferus	D																	
I. noduliferus	E																	
I. sulcatus subsp.	*Idiognathoides sulcatus* subsp.		1/0															
I. sulcatus subsp.	M.T.S. F																	
I. sulcatus subsp.	G																	
I. sulcatus subsp.	A																	
I. sulcatus subsp.	H																	
I. sulcatus subsp.	I																	
	Ozarkodina delicatula											1			1			
	Gnathodus bilineatus morpho. γ																	
	Idiognathoides sinuatus								0/3	1/0	27/28	48/68	75/74	16/22	5/4			1/0
	Idiognathoides sulcatus parvus												5/10	2/0	0/2			
	Adetognathus spathus				1/1							3/5	2/2				1/0	
	Neognathodus bassleri symmetricus					0/1	2/5	4/0	33/27	21/24		2/2						
	N. bassleri bassleri												20/20	7/9	1/3			1/1
	Idiognathodus sinuosis																	23/36
	Spathognathodus minutus															2		
	Spathognathodus n. sp.															1		
	Streptognathodus expansus																	

Group	Taxon	22-1	22-4	22-6	22-8	22-9	21-1	21-2	21-3	21-5	21-6	21-10	21-11	21-13	21-15	21-17	21-18	21-19
	FORMATION	HALE / BLOYD					BLOYD											
	MEMBER																	
	Cavusgnathus unicornis																	
	Cavusgnathus altus s.l.																	
	Gnathodus commutatus subsp.																	
	Rhachistognathus primus																	
	Adetognathus lautus	2/0	3/0	2/1	2/1	1/0	4/5	2/4	8/9	2/3	3/1	1/2	1/0	0/1	2/0	5/11	0/1	1/0
I. noduliferus	*Idiognathoides noduliferus*																	
I. noduliferus	M.T.S. B																	
I. noduliferus	C																	
I. noduliferus	A																	
I. noduliferus	D																	
I. noduliferus	E																	
I. sulcatus subsp.	*Idiognathoides sulcatus* subsp.																	
I. sulcatus subsp.	M.T.S. F																	
I. sulcatus subsp.	G																	
I. sulcatus subsp.	A																	
I. sulcatus subsp.	H																	
I. sulcatus subsp.	I																	
	Ozarkodina delicatula																	
	Gnathodus bilineatus morpho. γ																	
	Idiognathoides sinuatus				1/1		1/4	12/15	5/7	1/4						46/49		4/6
	Idiognathoides sulcatus parvus				0/1													
	Adetognathus spathus					1/0										3/0		
	Neognathodus bassleri symmetricus	1/0	1/2	2/0			2/5	2/1										
	N. bassleri bassleri																	
	Idiognathodus sinuosis														0/1	2/4		1/0
	Spathognathodus minutus			1														
	Spathognathodus n. sp.																	
	Streptognathodus expansus															1/0		

Figure 27. Distribution of conodont taxa in sampled Localities 16, 18, 21, 22, 24, and M-25. $\frac{0}{0}$ = **number of left specimens above number**

Locality M-51

TAXON	2	5	6	7	9	10	11	12	13	14	15	17	18	19	20	21	23	25	27	28	29	30
FORMATION	(?) PITKIN	(?) PITKIN	HALE	HALE	HALE	HALE	HALE	HALE	HALE	HALE	HALE	HALE ?	BLOYD	BLOYD	BLOYD	BLOYD	BLOYD	BLOYD	BLOYD	BLOYD	BLOYD	BLOYD
Cavusgnathus altus s.l.	0/3																					
Spathognathodus cristula	1																					
Gnathodus bilineatus morpho. ?			0/1																			
G. bilineatus morphotype α																						
G. bilineatus morphotype β				1/0																		
Gnathodus commutatus subsp.			1/0																			
Adetognathus lautus		1/1	1/1	2/2	6/6	1/0	6/3	2/1	12/16	4/3	9/11	2/3	4/1	2/1	6/11	0/1	2/0	0/5		4/5	10/7	13/12
Adetognathus spathus								0/1	0/1						1/0							0/1
I. noduliferus: *Idiognathoides noduliferus*																						
I. noduliferus: M.T.S. B				0/1	0/1																	
I. noduliferus: M.T.S. C			1/0	1/0																		
I. noduliferus: M.T.S. A			1/0																			
I. noduliferus: M.T.S. E												2/0										
I. sulcatus subsp.: *Idiognathoides sulcatus* subsp.							0/1															
I. sulcatus subsp.: M.T.S. F						0/1																
I. sulcatus subsp.: M.T.S. G																						
I. sulcatus subsp.: M.T.S. A																						
I. sulcatus subsp.: M.T.S. H																						
I. sulcatus subsp.: M.T.S. I			1/0	2/1																		
Rhachistognathus muricatus																						
Idiognathoides macer						0/1																
Neognathodus bassleri symmetricus						0/1		1/2	4/2	1/0	7/8	7/7			0/1							
Idiognathoides sinuatus							0/1		0/2		1/0				3/3					8/3		
Neognathodus bassleri bassleri																	1/0	1/0				
Idiognathoides sulcatus parvus																						
Idiognathodus sinuosis																				2/2	5/5	7/11
Spathognathodus minutus												2										
Ozarkodina sp.									5		1		1	1				1				
Ozarkodina delicatula																						
Spathognathodus n.s.p.																						
Idiognathoides convexus																					3/4	10/8

Locality 20

TAXON	22	21	20	19	18	17	16	15	14	13	12	11	9	8	7	6	5	4	2	1
FORMATION	HALE	HALE	HALE	HALE	HALE	HALE	HALE	HALE	HALE	HALE	HALE	HALE	BLOYD	BLOYD	BLOYD	BLOYD	BLOYD	BLOYD	BLOYD	BLOYD
Cavusgnathus altus s.l.					0/1															
Spathognathodus cristula																				
Gnathodus bilineatus morpho. ?			0/1		0/1															
G. bilineatus morphotype α					1/0															
G. bilineatus morphotype β																				
Gnathodus commutatus subsp.			1/0																	
Adetognathus lautus		3/3	2/0	1/0	5/6	3/12	2/1			0/1	4/1	0/1	2/3	0/4	11/7	4/2	2/7		0/2	1/0
Adetognathus spathus		2/1																		
I. noduliferus: *Idiognathoides noduliferus*			1/0																	
I. noduliferus: M.T.S. B		0/1																		
I. noduliferus: M.T.S. C	1/0	1/0																		
I. noduliferus: M.T.S. A																				
I. noduliferus: M.T.S. E																				
I. sulcatus subsp.: *Idiognathoides sulcatus* subsp.																				
I. sulcatus subsp.: M.T.S. F		0/1																		
I. sulcatus subsp.: M.T.S. G		1/0																		
I. sulcatus subsp.: M.T.S. A																				
I. sulcatus subsp.: M.T.S. H		0/1	0/1																	
I. sulcatus subsp.: M.T.S. I		2/0																		
Rhachistognathus muricatus		1/0																		
Idiognathoides macer																				
Neognathodus bassleri symmetricus					0/1	4/7	8/2	1/0	1/0		2/0									
Idiognathoides sinuatus														4/2	1/2	34/27	40/36	0/1		
Neognathodus bassleri bassleri																2/2	0/2			
Idiognathoides sulcatus parvus																8/7	24/21			
Idiognathodus sinuosis																				
Spathognathodus minutus															2					
Ozarkodina sp.						2									3					
Ozarkodina delicatula														1						
Spathognathodus n.s.p.															2					1
Idiognathoides convexus																				

Figure 28. Distribution of conodont taxa in sampled Localities 20 and M-51. $\frac{0}{0}$ **= number of left specimens above number of right specimens.**

Figure 29. Distribution of conodont taxa in sampled Locality 23. $\frac{0}{1}$ = number of left specimens above number of right specimens.

Formations: M-5 — PITKIN / HALE ? / BLOYD; M-28 — PITKIN / HALE ? / BLOYD; M-64 — PITKIN / HALE / ? BLOYD

Locality M-5

TAXON / SAMP.	1	2	6	7	8	10	11	13
Cavusgnathus altus s. l.	0/6							
Cavusgnathus unicornis	0/5							
Cavusgnathus navicuius								
Gnathodus commutatus subsp.								
Spathognathodus cristula								
Gnathodus bilineatus morphotype ?								
G. bilineatus morphotype α								
morphotype β								
Geniculatus claviger								
Rhachistognathus muricatus	1/0							
Adetognathus lautus		11/8	5/7	1/4	0/1	3/12	3/4	16/16
I. nodu. M.T.S. B								
I. nodu. M.T.S. D								
I. sulc. subsp. *Idiognathoides sulcatus subsp.*								
I. sulc. subsp. M.T.S. G								
I. sulc. subsp. M.T.S. I				1/0				
Idiognathoides sinuatus		1/1	1/0	1/1	3/2	21/13	3/3	
Neognathodus bassleri symmetricus		27/28	3/3	1/4				
Neognathodus bassleri bassleri						20/21	7/5	
Idiognathoides sulcatus parvus								
Idiognathodus sinuosis								91/93
Streptognathodus expansus								1/0
Streptognathodus sp.								1/1
Spathognathodus n. sp.					3			
Ozarkodina delicatula								
Spathognathodus minutus	1		1					
Adetognathus spathus		0/1						4/6
Ozarkodina sp.								
Neoprioniodus sp.			0/1	0/1				
Synprioniodina sp.								

Locality M-28

TAXON / SAMP.	1	3	5	6	7	9	10	15	16	17	18	19	20	21	22
Cavusgnathus altus s. l.															
Cavusgnathus unicornis															
Cavusgnathus navicuius															
Gnathodus commutatus subsp.															
Spathognathodus cristula	1														
Gnathodus bilineatus morphotype ?															
G. bilineatus morphotype α															
morphotype β															
Geniculatus claviger															
Rhachistognathus muricatus															
Adetognathus lautus		0/1		1/4	1/1	0/1	0/1	0/1			4/3	1/1	0/1	0/1	
I. nodu. M.T.S. B															
I. nodu. M.T.S. D			0/1												
I. sulc. subsp. *Idiognathoides sulcatus subsp.*															
I. sulc. subsp. M.T.S. G															
I. sulc. subsp. M.T.S. I															
Idiognathoides sinuatus					7/4				0/1	15/16	43/51	55/55	6/4		
Neognathodus bassleri symmetricus															
Neognathodus bassleri bassleri					0/1										
Idiognathoides sulcatus parvus					0/1										
Idiognathodus sinuosis										5/11	38/25	19/25	0/1		0/1
Streptognathodus expansus										1/0		2/0			
Streptognathodus sp.															
Spathognathodus n. sp.															
Ozarkodina delicatula		1										6			
Spathognathodus minutus									1						1
Adetognathus spathus												2/2			
Ozarkodina sp.											1				
Neoprioniodus sp.	1/0														
Synprioniodina sp.												3			

Locality M-64

TAXON / SAMP.	1	3	4	5	6	7	8	9	10	11	12	14	15	16	17	19	24
Cavusgnathus altus s. l.	0/2		0/1														
Cavusgnathus unicornis	0/6	0/1	0/2														
Cavusgnathus navicuius		0/1	0/7														
Gnathodus commutatus subsp.	0/1		4/0	1/0						1/0							
Spathognathodus cristula	1	1															
Gnathodus bilineatus morphotype ?		2/0	4/1					0/1									
G. bilineatus morphotype α			2/2														
morphotype β			7/5	3/0													
Geniculatus claviger			2														
Rhachistognathus muricatus			0/1														
Adetognathus lautus		1/0	2/0	3/1	10/14	3/6	20/25	3/1	2/1	1/2	3/1	3/4	0/2	12/8	2/0	1/3	1/0
I. nodu. M.T.S. B					0/1												
I. nodu. M.T.S. D																	
I. sulc. subsp. *Idiognathoides sulcatus subsp.*					1/0		1/0										
I. sulc. subsp. M.T.S. G					1/0												
I. sulc. subsp. M.T.S. I							0/2										
Idiognathoides sinuatus						0/2		1/1				5/6		1/0	1/0	2/1	5/5
Neognathodus bassleri symmetricus							4/1	1/1		0/1	9/11						
Neognathodus bassleri bassleri												0/1			0/1	1/0	
Idiognathoides sulcatus parvus																	
Idiognathodus sinuosis																	0/1
Streptognathodus expansus																	
Streptognathodus sp.																	
Spathognathodus n. sp.																	
Ozarkodina delicatula														1	1		
Spathognathodus minutus																	
Adetognathus spathus						0/1	9/6										
Ozarkodina sp.								1						2	1		
Neoprioniodus sp.			0/1		0/1												
Synprioniodina sp.															1		

Figure 30. Distribution of conodont taxa in sampled Localities M-5, M-28, and M-64. $\frac{0}{0}$ **= number of left specimens above number of right specimens.**

FORMATION	HALE ?														BLOYD													
LOC.	M–26														M–42																
TAXON \ SAMPLE	1	3	4	5	6	9	10	11	12	14	15	19	20	21	1	7	8	9	10	11	12	13	14	15	16	17	18	19	20	21
Adetognathus lautus	14/12	1/2	0/1	0/1	2/1	2/7	3/1	6/4	0/1	2/4	1/1	4/4	1/1	2/0	16/17		0/1	1/0	5/5	3/2	3/2	0/3	1/2	1/1	0/5		4/4	6/2	1/1	1/2
I. noduliferus M.T.S. E	1/0				1/0									2/0	0/1		0/1							1/1					1/1	1/2
I. sulcatus subsp. M.T.S. F	0/1																								1/2					
Adetognathus spathus	2/2				1/0	1/0				2/1	2/1	2/1	1/1		51/66			1/1	5/7	0/1		0/1	3/4	3/10						
Idiognathoides sinuatus	0/1				3/1	3/1	2/0	25/18							0/10			5/7	0/1		0/1	0/1					22/17		
Neognathodus bassleri symmetricus					0/1	0/1																							
Neognathodus bassleri bassleri							1/1			1/0					163/161														
Idiognathodus sinuosis							1/6			47/46	1/4	4/4		2/0	0/3	0/1						0/1			13/7	1/7	1/1	1/0	
Streptognathodus expansus										3/0	3/0				0/3	2/0													
Streptognathodus sp.										0/1			0/1		1/0														
Idiognathoides convexus												2/3											0/1	5/7	22/17			
Idiognathoides cf. I. macer															0/2													
Cavusgnathus altus s.l.																												
Spathognathodus minutus																5								0/1		2		
Ozarkodina delicatula	1									1																		
Hibbardella sp.										1					1	1												
Ozarkodina sp.	3									1					1	1												
Neoprioniodus sp.	0/2									0/1			0/1												1/1	1/0		

Figure 31. Distribution of conodont taxa in sampled Localities M-26 and M-42. $\frac{0}{0}$ = number of left specimens above number of right specimens.

Form-Genus *Adetognathus* Lane, 1967

Type-species: *Cavusgnathus lautus* Gunnell, 1933, p. 286, Pl. 31, figs. 67, 68; Pl. 33, fig. 9 (left element). The lectotype, selected and refigured herein (Fig. 40: 9-11) is the specimen illustrated by Gunnell (1933, Pl. 31, figs. 67, 68 [=UM 502-4]). *Cavusgnathus gigantus* Gunnell (holotype illustrated by Gunnell, 1933, p. 286, Pl. 33, figs. 7, 8; =UM C515-5) and refigured herein (Fig. 38: 1-4) is considered of equal taxonomic significance for the form-generic concept.

Diagnosis

The form-genus *Adetognathus* includes species possessing Class II and Class IIIb symmetry (Lane, 1968), and the platform has a deep, median, longitudinal trough throughout its length. The fixed blade, if developed, is much shorter than the long free blade, and the blade rises in height anteriorly. No carina is developed.

Remarks

Gunnell's illustrations (1933, Pl. 31, figs. 67, 68) appear to be upper and lower views of the same specimen. According to Ethington (1968, written commun.), type numbers were assigned prior to the time Gunnell reposited his specimens in the University of Missouri collections, and one number was assigned to each view. The University of Missouri repository number UM 502-3 was assigned to Figure 67, but the slide is empty. The slide labeled UM 502-4 contains a specimen that appears to have been illustrated in two views (Gunnell, 1933, Pl. 31, figs. 67, 68 [=holotype]). Thus, the designation of specimen 502-3 as the holotype by Ellison (1941, p. 126) must be in error.

In lower view, species of *Adetognathus* display a large, subequally flared basal cavity under the platform which continues to the anterior tip of the free blade as a slit.

Adetognathus differs from *Cavusgnathus* by complete absence or weak development of a fixed blade and in symmetry characteristics. All known representatives of *Cavusgnathus* have a fixed blade equal to, or longer than, the free blade and display Class IV symmetry. Lateral profile of the oral margin of the free blade also serves to distinguish these genera. *Cavusgnathus* possesses a short free blade, and its apex is in the posterior half; whereas, excepting an abnormally large posteriormost denticle, the longer free blade of *Adetognathus* has its highest point in the anterior half. In this concept, the specimen described by Rhodes and others (1969, p. 84, Pl. 9, figs. 10a-10d) cannot be regarded as belonging to *Cavusgnathus*.

Lane (1967, p. 930) suggested *Adetognathus* was derived from *Spathognathodus* or *Cavusgnathus,* and the latter genus was considered the more likely progenitor.

Range

Adetognathus ranges from the uppermost Chesterian (Grove Church Shale) of the Illinois Basin into lower Permian strata (Ellison, 1941, p. 126).

Adetognathus lautus (Gunnell, 1933)

(Fig. 36: 17, 21, 22, 25-31; Fig. 38: 1-4, 6-8, 10-15, 20; Fig. 39: 1-3, 7-14)

Cavusgnathus lautus Gunnell, 1933, p. 286, Pl. 31, figs. 67, 68; Pl. 33, fig. 9; Webster, 1969, p. 28, Pl. 4, fig. 9a, 9b.

Cavusgnathus gigantus Gunnell, 1933, p. 286, Pl. 33, figs, 7, 8; Webster, 1969, Pl. 4, figs. 6a, 6b.

Cavusgnathus missouriensis Gunnell, 1933, p. 286, Pl. 33, figs, 10, 11.

Cavusgnathus lauta Gunnell. Ellison, 1941, p. 126, Pl. 21, figs, 47, 48.

Cavusgnathus giganta Gunnell. Ellison, 1941, p. 126, Pl. 21, figs. 44, 45, 49; Ellison and Graves, 1941, Pl. 3, fig. 3; Youngquist and Downs, 1949, p. 162, Pl. 30, figs. 18-20; McLaughlin, 1952, p. 620, Pl. 83, figs. 3, 4, 6, 7.

Cavusgnathus flexa Ellison, 1941, p. 126, Pl. 21, figs. 42, 43, 46.

Cavusgnathus regularis Youngquist and Miller. Stibane, 1967, p. 333, Taf. 35, figs. 8-11, 14-16 (? *non* Taf. 35, figs. 11, 12, 17-19 = *Adetognathus* sp.).

Cavusgnathus cf. *regularis* Youngquist and Miller. Stibane, 1967, p. 333, Taf. 35, figs. 6, 7.

Cavusgnathus unicornis Youngquist and Miller. Stibane, 1967, p. 333, Taf. 35, figs. 1-3, 5.

Cavusgnathus sp. A. Koike, 1967, p. 295, Pl. 1, figs. 4a, 4b (=form-species *A. gigantus*).

Cavusgnathus sp. Koike, 1967, Pl. 1, fig. 5 (=form-species *A. lautus*).

Adetognathus giganta (Gunnell). Lane, 1967, p. 931, Pl. 120, figs. 16, 18, 19; Pl. 121, figs. 8, 12, 13, 16.

Adetognathus lauta (Gunnell). Lane, 1967, p. 933, Pl. 121, figs. 1-3, 7, 10, 11, 15, 17 (*non* figs. 4, 5, 18 = *A. spathus*).

Adetognathus lautus (Gunnell). Dunn, 1970a, p. 327, Pl. 61, figs. 1, 4.

Adetognathus gigantus (Gunnell). Dunn, 1970a, p. 326, Pl. 61, figs. 2, 3.

Diagnosis

This species displays Class IIIb symmetry. The left and right conodont elements of *Adetognathus lautus* display those features characteristic of the form-species *Adetognathus lautus* and *Adetognathus gigantus*, respectively. (See Lane, 1967, p. 933 and 931, respectively, for diagnosis and description of both the left and right elements.)

Remarks

Lane (1967, 1968) suggested that the left and right elements of this species, which he referred to two different form-species, may have paired in the conodont-bearing organism. Similar abundance in samples and complementary single-sidedness of the two form-species, coupled with an apparent coincidence of ranges, lend support to Lane's suggestion. *Adetognathus lautus* is the type-species of *Adetognathus* (Lane, 1967, p. 930) and, thus, must serve as the name-bearer in our element-pair specific concept. A complete synonymy of our morphologic concept of the element-pair species *A. lautus* is outlined above. However, the recognition of *Adetognathus lautus* as a Class IIIb symmetry species depends heavily on presentation of numerical occurrence charts and associated range data. Since most previous authors on Pennsylvanian conodonts have not recorded numerical and range information, our synonymy must be utilized with some caution.

Webster (1969, p. 26) regards *Cavusgnathus unicornis* Youngquist and Miller as a junior synonym of the form-species *Adetognathus gigantus*. We believe there is no phylogenetic relation between these two form-species, and their morphologic similarity reflects homeomorphy (see Lane, 1967, p. 932, for morphologic distinctions). Furthermore, Webster's (1969) apparent disparity in ranges of the two form-species composing *Adetognathus lautus* reflects his particular taxonomic concept (that is, *A. gigantus = A. gigantus + C. unicornis*), which is not accepted herein.

Ancestral relations of *Adetognathus lautus* are still unclear. However, additional material suggesting transitional stages from *A. unicornis* (Class II symmetry) via *A. spathus* (Dunn) to *A. lautus* (Class IIIb) is present in our collections. Recognition of the conodont paired-species, *A. lautus*, provides a stronger argument for the evolution of the Pennsylvanian-Permian adetognathids from *A. unicornis*, as suggested by Lane (1967, p. 932-934).

Occurrence

Adetognathus lautus first appears in the Target Limestone Lentil of the Springer Formation and ranges through the Academy Church Shale Member and to the uppermost Primrose Sandstone sampled in southern Oklahoma. The species first appears at the base of the Hale Formation in northeastern Oklahoma and northwestern Arkansas. *A. lautus* ranges to the top of the Bloyd Formation in both areas.

Material

Lefts: 804; rights: 847.

Types

SUI 33630-33633, 33712-33716, 33718-33722.

Adetognathus spathus (Dunn, 1966)

(Fig. 38: 5, 9, 16-19; Fig. 40: 4-6)

Cavusgnathus spatha (Dunn), 1966, p. 1297, Pl. 157, figs. 3, 7, 8; Webster, 1969, p. 28, Pl. 4, figs. 1, 4, 5.
Adetognathus lauta (Gunnell). Lane, p. 933, Pl. 121, figs. 4, 5, 18 (*non* figs. 1-3, 7, 10, 11, 14, 15, 17 = *A. lautus*).
Streptognathodus lanceolatus Webster, p. 47, Pl. 6, figs. 14, 15.
Adetognathus spathus (Dunn). Dunn, 1970a, p. 327, Pl. 61, figs. 11-13.

Diagnosis

Specimens of this species display Class IIIb symmetry. A denticulate posterior process, which is usually continuous with the outer margin, is developed.

Description

See Dunn (1966, p. 1297) for the description.

Remarks

Adetognathus spathus displays Class IIIb symmetry and seemingly represents a transitional stage between *A. unicornis* and *A. lautus*. However, some specimens we refer to *Rhachistognathus primus* Dunn in our collections strongly resemble *A. spathus* and conceivably are related to the latter species. *A. spathus* may have originated from both *Adetognathus unicornis* and *Rhachistognathus muricatus*. Further study is necessary to clarify the phylogenetic relations of this species.

Occurrence

Adetognathus spathus first occurs in lowest Morrowan strata and, in northeastern Oklahoma, ranges into strata that seemingly correlate with the Kessler Member of the Bloyd Formation (upper Morrowan). The species is known to range into Derryan age strata in Nevada (Lane and others, 1972). Specimens of the species have been recovered from Morrowan age rocks in southern Nevada (Webster, 1969, p. 29). See Dunn (1970a, p. 327) for other North American occurrences of the species.

Material

Lefts: 50; rights: 47.

Types

SUI 33711, 33717, 33723, 33724.

Adetognathus unicornis (Rexroad and Burton, 1961)

(Fig. 33: 14-18)

Taphrognathus varians Branson and Mehl. Cooper, 1947, p. 92, Pl. 20, figs. 14-16.
Streptognathodus unicornis Rexroad and Burton, 1961, p. 1157, Pl. 138, figs. 1-9;
  Collinson, and others, 1962, p. 27, Charts 1, 4; Dunn, 1965, p. 1149, Pl. 140,
  figs. 5, 6, 13, 14; Webster, 1969, p. 49, Pl. 4, figs. 13a, 13b.
Adetognathus unicornis (Rexroad and Burton). Lane, 1967, Pl. 119, figs. 16-20;
  Dunn, 1970a, p. 327, Pl. 61, figs. 20-22.

Diagnosis

This species displays Class II symmetry.

Left element: The free blade attaches to the left margin of the platform, and the blade-margin upper profile is gently convex outwardly. The posteriormost denticle of the free blade is abnormally large, and its right side may occupy a mid-platform position at the blade-platform junction. No fixed blade is developed.

Right element: Right-sided specimens in our collections assigned to *A. unicornis* closely compare as mirror-image representatives of the left element. The free blade attaches to the right margin of the platform; however, the posteriormost denticle is generally larger in right-sided forms.

Remarks

The position of the free blade and platform junction is marginal, although in oral view, the junction commonly appears to be central or subcentral due to the presence of the broadened abnormally large denticle at the posterior end of the free blade. The continuation of the blade with the outer platform can be readily seen when comparing inner and outer lateral views.

Lane (1967, p. 931) discussed the reasoning involved in transferring *Streptognathodus unicornis* Rexroad and Burton to *Adetognathus*. He also suggested that the form-species *A. unicornis* gave rise to all Pennsylvanian form-species referable to *Adetognathus*. *Adetognathus unicornis* is not related phylogenetically or morphologically to *Streptognathodus*. The latter genus possesses the gnathodid-type basal cavity, whereas *Adetognathus unicornis* displays the cavusgnathid-type basal cavity (see Lane, 1967, p. 937, for the distinctions between the cavusgnathid and gnathodid basal cavity) which precludes a direct phylogenetic relation. Lane (1967, p. 931) summarized other morphologic distinctions that set *A. unicornis* apart from the *Streptognathodus*. Thus, the concept that the *Taphrognathus-Cavusgnathus* lineage gave rise to *Streptognathodus* is not accepted herein.

Occurrence

Adetognathus unicornis is known to occur in the Grove Church Shale (=Kinkaid D) of the Illinois Basin (Cooper, 1947; Rexroad and Burton, 1961), in the Indian Springs and Bird Spring Formations of southern Nevada (Dunn, 1966, 1970a; Webster, 1969), and in the upper 15 ft of the Pitkin Limestone in Van Buren County, Arkansas (Lane, 1967). In southern Oklahoma, *A. unicornis* occurs in the Tiff Member and uppermost shales of the Goddard Formation, and in the Rod Club Member of the Springer Formation.

Material

Lefts: 33; rights: 29.

Types

SUI 33628, 33629.

Form-genus *Cavusgnathus* Harris and Hollingsworth, 1933

Type-species: *Cavusgnathus altus* Harris and Hollingsworth, 1933, p. 201, Pl. 1, figs. 10a, 10b.

Diagnosis

The members of this form-genus display Class IV symmetry. About three-quarters of the blade is fixed to the platform by virtue of the anterior extension of the left platform margin. The free blade is short and composes about one-quarter of the entire blade. Generally, the upper profile of the blade rises in height posteriorly.

Description

For description see Diagnosis of Thompson and Goebel (1968, p. 21).

Remarks

Distinguishing characteristics between *Adetognathus* and *Cavusgnathus* are discussed under the remarks of *Adetognathus*.

Cavusgnathus is separable into two broad morphological groups that are informally recognized herein as the *C. altus* group and the *C. naviculus* group. Members of the *C. altus* group are characterized by a deeply excavated, broad, U-shaped longitudinal trough, whereas specimens belonging to the *C. naviculus* group are typified by a filled-in trough region and occasionally a narrow, V-shaped trough that is restricted to the anterior part of the platform.

As explained under the remarks of the description of *Adetognathus*, the specimen figured and described by Rhodes and others (1969, p. 84, Pl. 9, figs. 10a-10c) is not regarded herein as belonging to *Cavusgnathus*.

Range

Cavusgnathus ranges from the middle part of the St. Louis Formation (Upper Meramec; Rexroad and Collinson, 1963) to the top of the Grove Church Shale (upper Chesterian; Rexroad and Burton, 1961).

Cavusgnathus naviculus (Hinde, 1900)

(Fig. 32: 6, 8, 10, 14-18)

Polygnathus navicula Hinde, 1900, p. 342, Pl. 9, fig. 5; Holmes, 1928, p. 18, Pl. 7, fig. 14.

Cavusgnathus cristata Branson and Mehl. Bischoff, 1957, p. 19, Pl. 2, figs. 7a, 7b.
Cavusgnathus inflexa Clarke, 1960, p. 23, Pl. 3, figs. 17, 19.
Cavusgnathus navicula (Hinde). Clarke, 1960, p. 23, Pl. 4, figs. 1–3; Rexroad and Burton, 1961, p. 1151, Pl. 139, figs. 4–13; Rexroad and Nicoll, 1965, p. 17, Pl. 1, figs. 24, 25; Higgins and Bouckaert, 1968, p. 29, Pl. 2, figs. 7, 8.
Cavusgnathus unicornis Youngquist and Miller. Higgins, 1961, Pl. 10, fig. 3.
Cavusgnathus naviculus (Hinde). ?Dunn, 1965, p. 1147, Pl. 140, figs. 10, 11; Rhodes and others, 1969, p. 81, Pl. 13, figs. 12a–12d, Pl. 14, figs. 1a–1d, 4a–4d; Webster, 1969, p. 28, Pl. 4, fig. 3; Dunn, 1970a, p. 329, Pl. 61, fig. 24 (? figs. 23, 25).

Diagnosis

The broad, massive platform is flat across the upper surface, with only a very narrow, slitlike trough which is generally restricted to the anterior one-half of the platform.

Description

The free blade attaches to the right margin, forming a straight to slightly sinuous right margin in upper view. The broad margins have coarse transverse ribs and random nodes. In lateral view, the upper profile of the blade is marked by a large prominent posteriormost denticle which is directed vertically or slightly inclined toward the posterior. From the tip of the main denticle, the profile rapidly descends, almost linearly, to the lower, anteriormost point of the blade. A median carina commonly is developed on the posterior one-quarter to one-half of the platform.

Remarks

Specimens departing somewhat from the typical *Cavusgnathus naviculus* morphology are illustrated (Fig. 32: 10, 17) from our collections. Both specimens are long, slender, graceful elements unlike the massive, broad *naviculus*-type. In addition, the lateral view (Fig. 32: 16) of Figure 32: 17 shows an inferior denticle developed posterior to the highest denticle of the blade, a feature not seen in previously figured specimens of *C. naviculus*. The specimen in Figure 32: 10 does possess the characteristic *C. naviculus* blade as illustrated (Fig. 32: 8). Both specimens figured (Fig. 32: 10, 17), and others like them in our collections, are assigned to *C. naviculus* on the basis of the filled-in trough region, coarse transverse ribbing and node development on the broad margins, development of a posterior, median carina, and similar or slightly modified denticulation of the posterior part of the free blade.

We feel that inclusion of these variant specimens in *C. naviculus* modifies the basic concept of morphology; consequently, we consider the variant elements, together with the more robust forms, as comprising the *C. naviculus*-group on trough development.

Occurrence

Cavusgnathus naviculus has been reported from the Menard (Rexroad and Nicoll, 1965) through Grove Church strata (Cooper, 1947; Rexroad and Burton, 1961) in the Illinois Basin, and from the Mississippian portion of the Bird Spring Formation Clark County, Nevada (Dunn, 1965). Lane (1967) reported *C. naviculus* from the upper 15 ft of the Pitkin Limestone in Van Buren County, Arkansas. Clarke (1960) reported the species from the *Posidonia* band below the Top Hosie Limestone, near Kilsyth, the Lower Limestone at Ponniel Water, Douglas, and the Hawthorn Limestone, Glenbuck, Muirkirk, Scotland. Higgins (1961) recovered *C. naviculus* from rocks of the *Eumorphoceras* aff. *pseudobilingue* Marine Band of North Staffordshire, England. Rhodes and others (1969) reported the species from their samples "North Crop 3D22" and "Avon Gorge D26."

In southern Oklahoma, *Cavusgnathus naviculus* ranges from the lower shales of the Goddard Formation, below the Redoak Hollow Member, to near the top of the formation. The species is known to occur in the top of the Pitkin Limestone in northeastern Oklahoma, and in some cases, it has been reworked into the base of the Hale Formation.

Material

Twenty-five right-sided elements with indeterminate curvature.

Types

SUI 33616–33618.

Cavusgnathus unicornis Youngquist and Miller, 1949

Cavusgnathus unicornis Youngquist and Miller, 1949, p. 619, Pl. 101, figs. 18–23; Rexroad, 1957, p. 17, Pl. 1, fig. 7; Rexroad, 1958, p. 17, Pl. 1, figs. 6–11; Rexroad and Burton, 1961, p. 1152, Pl. 138, figs. 10–12; Rexroad and Collinson, 1963, p. 9, Pl. 1, figs. 26, 27; Rexroad and Furnish, 1964, p. 670, Pl. 111, fig. 6; Rexroad and Nicoll, 1965, p. 18, Pl. 1, figs. 18–20; Globensky, 1967, p. 439, Pl. 57, figs. 5, 14; Koike, 1967, p. 294, Pl. 1, figs. 2, 3; Thompson and Goebel, 1968, p. 23, Pl. 1, figs. 2, 5, 6, 8; Rhodes and others, 1969, p. 82, Pl. 31, figs. 13a, 13b; Dunn, 1970a, p. 329, Pl. 61, fig. 29.
[non] *Cavusgnathus unicornis* (Youngquist and Miller). Stibane, 1967, p. 333, Pl. 35, figs. 1–5 (=form-species *Adetognathus gigantus)*.

Remarks

Cavusgnathus unicornis is considered, along with *C. altus* and *C. regularis*, as comprising the *C. altus*-group of cavusgnathids. Deep, U-shaped development of the median, longitudinal trough in the platform surface characterizes this group.

Occurrences

Cavusgnathus unicornis ranges from the middle part of the St. Louis Formation (upper Meramec) through the Grove Church Shale (upper Chester) in the Illinois Basin (Rexroad, 1957, 1958; Rexroad and Burton, 1961; Rexroad and Collinson, 1963; Rexroad and Nicoll, 1965). Specimens referable to this species have also been recovered from the Pella Formation of Iowa (Youngquist and Miller, 1949; Rexroad and Furnish, 1964), the Windsor Group of the Atlantic Provinces of Canada (Globensky, 1967), and the Nagoe Formation, Okayama Prefecture, southwest Japan (Koike, 1967). Thompson and Goebel (1968) report *C. unicornis* from the St. Louis and Ste. Genevieve Limestones of Kansas, and Rhodes and others (1969) reported the species to range from the upper part of the C Zone into the D Zone of the Avon Gorge and the north crop of the South Wales coalfield.

Cavusgnathus unicornis occurs just above the Caney-Goddard contact and ranges through the top of the Tiff Member of the Goddard Formation in the Ardmore Basin, southern Oklahoma. In northeastern Oklahoma and Arkansas, the species is known to occur in the Pitkin Limestone and is also commonly reworked into the base of the Hale Formation.

Material

Thirty-three right-sided specimens with indeterminate curvature.

Form-genus *Gnathodus* Pander, 1856

Gnathodus Pander, 1856, p. 33.
Type-species: *Gnathodus mosquensis* Pander, 1856, p. 34, Pl. 2, figs. 10a–10c (original designation).

Diagnosis

The upper surface of the platform is characterized by a median nodose or ribbed carina that extends to the posterior tip. In lower view, a large, flared, subsymmetrical to asymmetrical basal cavity is situated at the posterior end of the element.

Description

Gnathodus is a platform form-genus that displays Class II symmetry. The carina may or may not be flanked by shallow adcarinal troughs. The margins of the platform may be smooth or weakly to strongly ornamented by nodes and transverse ridges. The cavity may flare beyond the posterior end of the platform.

Remarks

Gnathodus can be differentiated from *Neognathodus* in that the former form-genus possesses an uninterrupted median carina that joins the posterior tip of the platform, and weak, unequally developed adcarinal troughs. Except for late forms, *Neognathodus* has a polygnathid appearance in upper view. We further distinguish the two form-genera phylogenetically. *Gnathodus* ancestry lies in the Devonian spathognathodids, while *Neognathodus* is derived through *Idiognathoides*.

Range

Uppermost Upper Devonian (Louisiana Limestone; Scott and Collinson, 1961) to upper Chesterian (Dunn, 1965).

Gnathodus bilineatus (Roundy, 1926)

(Fig. 32: 1-5, 7, 9, 11-13; Fig. 33: 11-13, 19-23, 25, 28-32; Fig. 34: 13-26; Fig. 40: 27)

Remarks

Hass (1953, p. 79) determined the form-species *Polygnathus bilineatus* Roundy to be conspecific with the form-species *Polygnathus texanus* Roundy and synonymized the two under the form-species *Gnathodus bilineatus* (holotype = *P. bilineatus* Roundy, 1926, Pl. 3, figs. 10a-10c). His description, however, more closely characterized the holotype of *Polygnathus texanus* than that of *P. bilineatus*. Hass' specific description and reillustration (1953, Pl. 14, figs. 26, 28) of Roundy's types, and additional specimens figured by Hass (1953, Pl. 14, figs. 25, 27, 28), which are again reillustrated herein (Fig. 33: 14, 16-20, 22, 25, 26), seem to have created a concept of "*bilineatus*," evident from later literature, which does not revolve about the holotype of the form-species *Polygnathus bilineatus* Roundy. Rather, the concept revolves around the larger, more visually arresting holotype of *Polygnathus texanus* Roundy. This "*bilineatus*-concept" emphasized the development of a concentric ridge pattern on the oral surface of the outer, lobate platform (*sensu P. texanus* Roundy holotype) rather than an irregularity in size and arrangement of node development on the same structure (*sensu P. bilineatus* Roundy holotype).

Specimens in our collection assigned to this species demonstrate a wide range of size and morphologic variation. The small illustrated specimens (Fig. 33: 11-13, 19-21, 30) show generalized features of free-blade development, inner platform development, and semicircular to triangular lobate platform shape, to be identical in all small forms. These specimens differ, however, in outer lobate platform ornament, which is seen to be of four basic types: (1) irregularly sized and arranged nodes covering the entire lobate platform (Fig. 33: 12, 25, 30; =morphotype α); (2) concentric nodose ridges paralleling the outer margin of the lobate platform (Fig. 33: 13, 23; =morphotype β); (3) anteriorly restricted node development that originates next to the blade and radiates toward the outer margin of the lobate platform (Fig. 33: 11, 32; =morphotype γ); (4) a single nodose ridge that passes around the outer margin of the lobate platform resulting in a circular pit in the center of the platform (Fig. 33: 20, 31; =morphotype δ). The arrangement, or pattern, of ornamentation developed on the lobate platform is established in the smallest specimens ("juveniles") and is maintained through the largest ("gerontic") elements. Except for this lobate platform ornamentation, every other morphologic feature (that is, inner margin, outer platform shape, depth of adcarinal trough) undergoes multidirectional variability through ontogeny without respect to outer platform ornament type. For example (Fig. 34: 14, 15), the reillustrated specimens of *Polygnathus texanus* Roundy possess the morphotype β lobate platform development, but the inner platform margin of one (Fig. 34: 14) is more sinuous than the other (Fig. 34: 15). The carina of one (Fig. 34: 15) shows stronger transverse-ribbed ornamentation than the other (Fig. 34: 14). These same two morphologic differences can be seen on any other morphotype. We believe all morphologic features, excepting lobate platform ornament, are too variable to distinguish these forms specifically. Lobate platform ornament patterns recognized herein, and informally distinguished nomenclatorially by Greek letters, may be specifically diagnostic. However, we hesitate to recognize them as such until these patterns can be studied over the stratigraphic range of *Gnathodus bilineatus*. In our own collections these morphotypes do not seem to be of great stratigraphic import, and do display some intergradation.

Gnathodus bilineatus Morphotype α

(Fig. 32: 5, 7, 9; Fig. 33: 12, 25, 30; Fig. 34: 10, 13, 16–18, 21; Fig. 40: 27)

Polygnathus bilineatus Roundy, 1926, p. 13, Pl. 3, figs. 10a–10c.
Gnathodus pustulosus Branson and Mehl, 1941a, p. 172, Pl. 5, figs. 33, 35–38 (*non* figs. 32, 34, 39).
Gnathodus liratus Youngquist and Miller, 1949, p. 619, Pl. 101, figs. 15–17.
Gnathodus bilineatus (Roundy). Hass, 1953, p. 78, Pl. 14, figs. 25, 26 (*non* figs. 27, 28, 29); Voges, 1959, p. 282, Pl. 33, figs. 29, 30? (*non* fig. 28); Elias, 1956a, Pl. 3, fig. 40; Globensky, 1967, p. 440, Pl. 58, fig. 9 (*non* fig. 10 = indet. morphotype).
Gnathodus modocensis Rexroad, 1957, p. 30, Pl. 1, fig. 17a, 17b (*non* figs. 15, 16 = indet. morphotype); Rexroad, 1958, p. 17, Pl. 1, fig. 2 (*non* fig. 1 = indet. morphotype).
Gnathodus bilineatus bilineatus (Roundy). Bischoff, 1957, p. 21, Pl. 3, figs. 11, 20 (*non* figs. 15–19); Pl. 4, fig. 1.
Gnathodus streptognathoides Elias, 1956a, p. 119, Pl. III, figs. 54, 57 (*non* figs. 55, 56 = indet. morphotype).
Gnathodus bransoni Elias, 1959, p. 12, Pl. 1, fig. 15 (redrawn from Branson and Mehl, 1941a, Pl. 5, fig. 38); fig. 13 (redrawn from Hass, 1951, Pl. 1, fig. 1).
Gnathodus bilineatus modocensis Rexroad and Furnish, 1964, p. 670, Pl. 111, fig. 5 (*non* fig. 4 = indet. morphotype).
Gnathodus bilineatus bollandensis Higgins and Bouckaert, 1968, p. 29, Pl. 2, figs. 10, 13; Pl. 3, figs. 4–8, 10.

Diagnosis

The outer lobate platform in all growth stages is ornamented by irregular sized and arranged nodes that cover the entire upper surface.

Remarks

This morphotype includes the holotype of the form-species *Polygnathus bilineatus* Roundy (USNM 115101).

Occurrence

Morphotype α ranges through the Goddard Shale and into the Rod Club Member of the Springer Formation in the Ardmore Basin. One specimen was found in the lower part of the Hale Formation in northeastern Oklahoma and is interpreted to be reworked.

The morphologic type is recognized in faunas illustrated from the Barnett Shale of Texas (Roundy, 1926), the Caney Shale of Oklahoma (Branson and Mehl, 1941a), the Pella Formation of Iowa (Youngquist and Miller, 1949; Rexroad and Furnish, 1964), and the upper faunal zone of the Barnett Shale (Hass, 1953). Elias (1956a) figured the morphotype from the lower part of the Goddard Shale of Oklahoma, and Globensky (1967) recovered it from the Windsor Group of the Maritime Provinces of Canada. Morphotype α occurs in the faunas illustrated from the Glen Dean (Rexroad, 1957) and Paint Creek (Rexroad, 1958) Formations of the Illinois Basin, the Stanley Shale of Oklahoma (Hass, 1951), and the Delaware Creek and Sand Branch Members of the Caney Formation from the Ada, Oklahoma, region (Elias, 1959). Bischoff (1957) reported the morphotype from the Iberger Kalk (*cu*III α), and Voges (1959) illustrated this form from *cu*II δ.

Material

Lefts: 125; rights: 156.

Types

SUI 33663, 33666, 33671, 33673; USNM 115100, 115101.

Gnathodus bilineatus Morphotype β

(Fig. 32: 2-4; Fig. 33: 13, 23; Fig. 34: 14, 15, 19, 20, 22-26)

Polygnathus texanus Roundy, 1926, p. 14, Pl. 3, figs. 13a, 13b.
Gnathodus pustulosus Branson and Mehl, 1941a, p. 172, Pl. 5, figs. 32, 34, 39.
Gnathodus bilineatus (Roundy). Hass, 1953, p. 78, Pl. 14, figs. 27, 28, 29; Voges, 1959, p. 282, Pl. 33, fig. 28; Higgins, 1961, Pl. 10, fig. 5; Dunn, 1970a, p. 330, Pl. 62, fig. 14.

Gnathodus multilineatus Elias, 1956a, p. 119, Fig. 49; Elias, 1959, p. 149, Figs. 26–28.

Gnathodus bilineatus bilineatus (Roundy). Bischoff, 1957, p. 21, Pl. 3, figs. 15–19; Wirth, 1967, p. 205, Pl. 19, figs. 6, 7a, 7b, 8, 9; Higgins and Bouckaert, 1968, p. 29, Pl. 3, fig. 9.

Gnathodus liratus (Youngquist and Miller). Elias, 1959, p. 148, Pl. 1, figs. 19, 20 (*non* fig. 21 = indet. morphotype).

Gnathodus minutus Elias, 1959, p. 148, Pl. 1, figs. 24, 25 (*non* figs. 22, 23 = indet. morphotype).

Gnathodus cf. *G. multilineatus* Elias, 1959, Pl. 1, fig. 29.

Gnathodus bransoni Elias, 1959, p. 147, Pl. 1, figs. 16, 17 (redrawn from Hass, 1953, Pl. 14, figs. 28, 29 (29 = holotype of *G. bransoni*); figs. 14, 15 (redrawn from Branson and Mehl, 1941a, Pl. 5, fig. 32); (*non* fig. 18 = indet. morphotype).

Gnathodus smithi Clarke, 1960, p. 26, Pl. 4, figs. 13, 14; Pl. 5, fig. 10 (*non* fig. 9 = indet. morphotype).

Diagnosis

The outer lobate platform in all growth stages is ornamented by concentric nodose ridges that parallel the outer margin of the platform. Elements possessing linear, nodose ridges that surmount the outer lobate platform parallel to the carina are also included.

Remarks

Morphotype β includes the holotype of the form-species *Polygnathus texanus* Roundy (USNM 115103).

Occurrence

Gnathodus bilineatus morphotype β has been illustrated from the Barnett Shale of Texas (Roundy, 1926; Hass, 1953), the Caney Shale of Oklahoma (Branson and Mehl, 1941a), the Sand Branch Member of the Caney Shale of the northern Arbuckles (Elias, 1956a, 1959), and the Stanley Shale of the Ouachita region (Elias, 1959). Higgins (1961) recorded the morphotype from the *Eumorphoceras* aff. *pseudobilingue* Marine Band in North Staffordshire, England, and Clarke (1960) reported it from the Upper Limestone Group and the Castlecary Limestone and Shales of the Midland Valley of Scotland. Morphotype β occurs in faunas from *cu*II δ through *cu*III γ strata of the Spanish West Pyrenees (Wirth, 1967), *cu*II δ and *cu*III α strata of the Sauerland in Germany (Voges, 1959), and *cu*III α through *cu*III γ strata in Germany (Bischoff, 1957).

Morphotype β was recovered from the middle Goddard Formation including the Tiff Member. The morphotype's occurrence in the lower part of the Hale Formation in northeastern Oklahoma and northwestern Arkansas is attributed to reworking.

Material

Lefts: 83; rights: 58.

Types

SUI 33662, 33667, 33670; USNM 115102–115104.

Gnathodus bilineatus Morphotype γ

(Fig. 32: 12; Fig. 33: 11, 22, 28, 29, 32)

Gnathodus bilineatus (Roundy). Dunn, 1970a, p. 330, Pl. 62, fig. 13.

Diagnosis

The outer lobate platform in all growth stages is ornamented by a radiating pattern of node development that originates next to the carina and radiates out to the outer margin of the platform. The ornament is generally confined to the anterior half of the lobate platform.

Occurrence

Gnathodus bilineatus morphotype γ was recovered from the middle Goddard Formation including the Tiff Member. The morphotype also was recovered from the top of the Pitkin Limestone in northeastern Oklahoma.

Material

Lefts: 55; rights: 51.

Types

SUI 33664, 33665, 33669, 33672.

Gnathodus bilineatus Morphotype δ

(Fig. 32: 1, 11, 13; Fig. 33: 19–21, 31)

Diagnosis

The outer lobate platform in all growth stages is ornamented by a single, narrowly or broadly nodose ridge that "fringes," or encircles the lobate platform at its margin, resulting in a circular pit which may or may not have nodes developed in it.

Occurrence

Gnathodus bilineatus morphotype δ was recovered from the Tiff Member of the Goddard Formation of the Ardmore Basin and the upper 15 ft of the Pitkin Limestone in north-central Arkansas. The morphotype's occurrence in the lower part of the Hale Formation in northeastern Oklahoma is attributed to reworking.

Material

Lefts: 9; rights: 12.

Types

SUI 33661, 33668, 33674.

Gnathodus commutatus commutatus (Branson and Mehl, 1941a)

(Fig. 37: 1-9; Fig. 40: 15-18, 23-26)

Spathognathodus commutatus Branson and Mehl, 1941a, p. 98, Pl. 19, figs. 1-4.
Gnathodus commutatus commutatus (Branson and Mehl). Bischoff, 1957, p. 23, Pl. 4, figs. 2-6, 15; Dunn, 1970a, p. 331, Pl. 62, figs. 11, 12.

Diagnosis

For diagnosis, see Bischoff (1957, p. 23).

Remarks

All recovered specimens referable to *Gnathodus commutatus* from Arkansas and Oklahoma are unornamented forms and thus are assigned to the nominate subspecies. Variability occurs in the breadth of the median carina, which varies from a knifelike nodose ridge to a carina of broad nodes or very short transverse ridges down the median axis of the platform. Generally, forms recovered from lower in the section display the former type of carina, while younger forms tend to develop a broader carina. However, broad carinal forms are present in our lowermost samples, and narrow carinal variants were recovered throughout the range of the species.

Occurrence

Gnathodus commutatus commutatus ranges from at least lowermost Goddard Shale through the Rod Club Member of the Springer Formation in southern Oklahoma. Specimens referable to this form-subspecies are present in the Fayetteville Shale and Pitkin Limestone in northwestern Arkansas and northeastern Oklahoma. The occurrence of *G. commutatus commutatus* in the Hale Formation in northwestern Arkansas and northeastern Oklahoma is attributed to reworking.

Material

Lefts: 246; rights: 281.

Types

SUI 33624–33627.

Gnathodus girtyi intermedius Globensky, 1967

(Fig. 33: 1–10, 24, 26, 27)

Gnathodus girtyi intermedius Globensky, 1967, p. 440, Pl. 58, figs. 11, 16–20.
Gnathodus? sp. Rexroad, 1958, p. 18, Pl. 1, figs. 3–5.

Diagnosis

Globensky's (1967, p. 440) description presents diagnostic characters of this form-subspecies. Additionally, the form-subspecies displays Class II symmetry, and the flaring of the basal cavity incorporates the upper, anterior part of the outer margin, forming a gently rounded "shoulder" that gradually slopes to the lower margin.

Remarks

Globensky's illustrated specimens do not include small forms, but the collections from southern Oklahoma include a range of sizes that lends support to Globensky's (1967, p. 440) suggestion that the form-subspecies was derived from *Gnathodus girtyi girtyi*.

The outer margin of the platform is poorly developed, with a few isolated nodes (Fig. 33: 2, 9). Larger specimens (Fig. 33: 4, 5, 26) show a gradual development of the outer margin into a nodose, continuous ridge. The basal cavity increasingly becomes strongly expanded through this ontogenetic series, especially to the outer lateral side.

A variation in upper profile of the free blade was noted and illustrated (Fig. 33: 1, 8) from our *G. girtyi intermedius* material. The upper profile gradually rises from the blade-platform junction to an apex at the largest denticle, which is positioned at the midpoint of blade length. A break in profile occurs anterior to the apex, and the anterior one-half of the blade is surmounted by very broad, low "sub-denticles." A small denticle near the upper margin, but on the leading edge of the blade, projects anteriorly. This upper outline is not seen on larger specimens of this form-subspecies in our collections, nor in the lateral view of the specimen illustrated by Globensky (1967, Pl. 58, fig. 17), and may be a "juvenile" character. However, larger collections over the stratigraphic interval yielding this form may produce larger specimens showing the same feature, in which case it would become subspecifically diagnostic.

The smaller illustrated specimens of the subspecies compare closely to those described and illustrated by Rexroad (1958, p. 18, Pl. 1, figs. 3-5) and are considered to be consubspecific.

Occurrence

Gnathodus girtyi intermedius has been reported from the Windsor Group of the Atlantic Provinces of Canada (Globensky, 1967) and from the Glen Dean Formation of the Illinois Basin (Rexroad, 1958).

The form-subspecies was recovered from the lower 125 ft (38.1 m) of the unnamed shale unit that overlies the "Overbrook" Sandstone Member of the Springer Formation in the Ardmore Basin.

Material

Lefts: 1; rights: 11.

Types

SUI 33656-33660

Form-genus *Idiognathodus* Gunnell, 1933

Type-species: *Idiognathodus claviformis* Gunnell, 1931, p. 249, Pl. 29, figs. 21, 22.

Diagnosis

The long free blade attaches in a median position and forms a short carina on the anterior part of the platform. The posterior part of the platform generally is transversely ornamented with ridges, and no trough is developed.

Description

The platform form-genus may display either Class II or Class IIIb symmetry. The carina is flanked on each side by one to three rostral ridges.

Nodose lobes are commonly well developed on the anterior inner margin and may be weakly to strongly developed on the anterior outer margin. The posterior half of the platform bears transverse nodes or ridges. In lower view, a large, deep, asymmetrical gnathodid-type basal cavity is developed.

Remarks

Idiognathodus Gunnell and *Streptognathodus* Stauffer and Plummer are differentiated solely on the presence of a median longitudinal trough that interrupts the posterior transverse ornament in the latter form-genus. Intergradations between these two form-genera are present in the Morrowan collections discussed herein. Furthermore, this intergradation seems to persist throughout the range of both form-genera. For the purpose of this report, both form-genera are provisionally retained because it is thought that a comprehensive study of faunas spanning the total vertical range of these two forms is necessary in order to establish their phylogenetic and morphologic relations.

Range

Idiognathodus is known to occur as low as the Woolsey Member of the Bloyd Formation, Morrowan (Lane, 1967), and to range at least to the top of the Pennsylvanian (Hass, 1962).

Idiognathodus klapperi n. sp.

(Fig. 42: 12–16)

Idiognathodus n. sp. A. Lane and Straka *in* Lane and others, 1971, Pl. 1, fig. 15.

Diagnosis

The form-species displays Class II symmetry. An inner anterior nodose lobe may be developed, but rostral ridges are absent or indistinct. The middle of the platform is ornamented with rows of nodes that radiate indistinctly from the posterior end of the carina and show a weak transverse lineation. The platform upper surface is flat and of the same height as the posterior end of the free blade.

Description

The form-species possesses a long free blade that meets the circular to oval platform in a median position. The carina extends about one-third the length of the platform. The posterior one-third of the platform may be ornamented with discontinuous transverse ridges that consist of discrete or fused nodes.

In lateral view, the free blade is highest anteriorly and descends in height posteriorly. The posterior tip of the upper surface drops vertically to meet the posterior tip of the basal cavity.

In lower view, a subtriangular gnathodid-type basal cavity is present.

Remarks

The rounded posterior platform margin, high flat upper surface, and radiate rows of nodes serve to distinguish the species from all other idiognathodids. Weak development of the inner anterior lobe, of the rostral ridges, and of the posterior transverse ridges in *Idiognathodus klapperi* differentiate it from *I. claviformis* Gunnell.

The form-species is named in honor of Dr. Gilbert Klapper.

Occurrence

Idiognathodus klapperi occurs in the caprock of the Baldwin coal (Dye Shale Member of the Bloyd Formation), Washington County, Arkansas.

Material

Lefts: 4; rights: 2.

Types

Holotype: SUI 33767; paratypes: SUI 33765, 33766.

Idiognathodus sinuosis Ellison and Graves, 1941

(Fig. 37: 10–13, 21; Fig. 42: 1–11; Fig. 43: 1–8, 10–15, 19, 20)

Idiognathodus sinuosis Ellison and Graves, 1941, p. 6, Pl. 3, fig. 22; Dunn, 1970a, p. 333, Pl. 63, figs. 3, 4.
Idiognathodus humerus Dunn, 1966, p. 1300, Pl. 158, figs. 6, 7; Dunn, 1970a, p. 333, Pl. 63, figs. 1, 2.

Diagnosis

The species displays Class IIIb symmetry. The left element is widest immediately posterior of the outer rostral ridge, and the posterior two-thirds of the platform is ornamented with anteriorly directed chevron-shaped transverse ridges. The right element is distinctly incurved, and maximum platform width is farther posterior than in the left element. The transverse ridges trend diagonally, with the inner ends extending farther anteriorly than the outer ends.

Description

Left element: The anterior median carina, extending about one-third of platform length, is paralleled by two to four rostral ridges comprised of fused or discrete nodes. The outer platform margin is widest immediately posterior of the outer rostral ridge. A nodose lobe is generally developed on the anterior inner margin, but it may be absent in very small forms. A weak nodose outer lobe is developed in larger specimens. The posterior two-thirds of the platform is ornamented with transverse ridges, which form an anteriorly directed chevron pattern centered on the trough. The shallow trough generally extends the length of the posterior platform.

Right element: This form is closely similar to the left element. However, maximum platform width is farther posterior than in the left element. The outer margin is formed as a continuation of the outer rostral ridge. Also, the platform is distinctly incurved, and the transverse ridges trend diagonally with the inner ends extending farther forward than the outer ends.

Remarks

Dunn (1966, p. 1301; 1970a, p. 333) noted that the form-species *Idiognathodus humerus* is left-sided. He also reported the direct association of an approximately equal number of right-sided forms thought to be conspecific with *Idiognathodus sinuosis* Ellison and Graves. If these observations are correct, then *I. sinuosis* is the valid name for the element-pair species.

In contrast to the form-species *Idiognathodus sinuosis*, the form-species *I. incurvus* Dunn has better developed inner and outer lobes, a wider platform, and almost directly transverse ridges. The form-species *Idiognathodus incurvus* is judged not to be synonymous with the right element of the element-pair species *I. sinuosis*.

The holotype of the form-species *I. humerus* (Dunn, 1966, Pl. 158, figs. 6, 7) is considered of equal taxonomic significance for the specific concept of the element-pair species *I. sinuosis*.

Gerontic specimens of the left element (Fig. 37: 6, 14) possess indistinct ornamentation. The transverse ridges on the posterior two-thirds of the platform tend to become discontinuous, with loss of the anteriorly directed chevron pattern.

Occurrence

Idiognathodus sinuosis is known to occur in the uppermost Primrose Member of the Golf Course Formation, Carter County, Oklahoma. It has also been retrieved from the Woolsey, Dye Shale, Kessler Limestone, and Trace Creek Shale Members of the Bloyd Formation in Washington County, Arkansas. The species also occurs in the Bloyd Formation of northeastern Oklahoma.

Material

Lefts: 309; rights: 363.

Types

SUI 33641, 33642, 33754–33764, 33774–33785.

Form-genus *Idiognathoides* Harris and Hollingsworth, 1933

Idiognathoides Harris and Hollingsworth, 1933, p. 201, Pl. 1, fig. 14; Hass, 1959, p. 379; Hass, 1962, p. W62; Lane, 1967, p. 936, Pl. 119, figs. 1-9, 12-15, Pl. 122, figs. 1, 2, 4-7, 9-11, Pl. 123, figs. 7-19; Igo and Koike, 1968, p. 28, Pl. III, figs. 7-10, 12 (*non* figs. 11a, 11b = ?*Gnathodus*); Higgins and Bouckaert, 1968, p. 38, Pl. 2, fig. 14, Pl. 4, figs. 3, 5-10, Pl. 5, figs. 9, 11, Pl. 6, figs. 1-6 (*non* figs. 7-12 = *Spathognathodus*); Webster, 1969, p. 37, Pl. 5, figs. 17, 18; Palmieri, 1969, p. 6, Pl. 2, figs. 4-9, Pl. 3, figs. 1-4, 7-12, 16, 17, 22-26, Pl. 4, figs. 6-13, 22-25; Dunn, 1970a, p. 334, Pl. 61, fig. 17, Pl. 63, figs. 14-18, 20-25, 28-30.

Polygnathodella Harlton, 1933, p. 15, Pl. 4, figs. 14a-14c; Branson and Mehl, 1941b, p. 103, Pl. 19, figs. 27, 28; Ellison and Graves, 1941, p. 8, Pl. 3, figs. 8-16; McLaughlin, 1952, p. 620, Pl. 83, figs. 1, 5; Fay, 1952, p. 149, fig. 9D; Clarke, 1960, p. 28, Pl. V, figs. 11-13, 16; Lindström, 1964, p. 173, fig. 63a; Wirth, 1967, p. 222, Pl. 20 figs. 12, 14; text Figure 12a-12d; Koike, 1967, p. 308, Pl. 3, figs. 1-5.

Declinognathodus Dunn, 1966, p. 1299, Pl. 158, figs. 4, 8; Dunn, 1970a, p. 330, Pl. 62, figs. 1, 2, 5-7.

Type-species: *Idiognathoides sinuata* Harris and Hollingsworth, 1933, p. 201, Pl. 1, fig. 14; also Lane, 1967, Pl. 123, fig. 8, for refigured holotype (left element). The lectotype of *Idiognathus corrugata* Harris and Hollingsworth, selected herein (= Harris and Hollingsworth, 1933, Pl. 1, figs. 8a, 8b) and refigured in this study (Fig. 41: 9; right element), is considered of equal taxonomic significance for the generic concept.

Diagnosis

The form-genus displays Class II or Class IIIb symmetry. In upper view, the elements possess a median trough of variable length. A blade that curves laterally to meet the anterior outer margin may be present on each element of the pair. In lower view, a large, deep, asymmetrical, gnathodid-type of basal cavity is developed.

Remarks

Lane (1967, p. 936) clarified the taxonomic status of *Idiognathoides* and summarized previous references to descriptions of the form-genus. Later, Lane (1968, p. 1260) suggested that individual elements of two form-species (*I. sinuatus*, only lefts; *I. corrugatus*, only rights) belonging to *Idiognathoides* formed an asymmetrical pair (Class IIIb) in the conodont-bearing organism. Faunas studied for the current investigation support the foregoing suggestion.

Range

Idiognathoides first appears at or near the base of the Morrowan and ranges into the Atokan.

Idiognathoides convexus (Ellison and Graves, 1941)

Polygnathodella convexa Ellison and Graves, 1941, p. 9, Pl. 3, figs. 10, 12, 16.
Idiognathoides convexus (Ellison and Graves) Webster, 1969, p. 37, Pl. 5, fig.
18; Lane and others, 1971, Pl. 1, figs. 17, 18.

Diagnosis

The species displays Class IIIb symmetry. The platform of the right element
is narrow, and the transverse-ridged ornamentation on the upper surface is convex
toward the posterior.

Description

Left element: The left element of *I. convexus* is much like the left element
of *I. sinuatus,* except that the central trough is generally longer and deeper in
I. convexus.
Right element: See Dunn (1970a, p. 334) for a description of the right element
of *I. convexus.*

Remarks

The specimens illustrated by Webster (1969, Pl. 5, fig. 17) and Dunn (1970a,
Pl. 63, fig. 20) are not considered as being within the concept of *I. convexus*
herein, because those specimens seem to be developing a centrally located blade.
They may represent specimens transitional into, or in juvenile stages of, *Idiognatho-*
dus. We have recovered forms that are very close to right elements of *I. convexus*
from our *N. bassleri symmetricus* unit in southern Nevada. Whether these specimens
represent an extension of the range of *I. convexus,* homeomorphs, or variants
of the right element of *I. sinuatus* cannot be answered at this time.

Occurrence

Idiognathoides convexus is known to occur in the Kessler Limestone Member
of the Bloyd Formation in northwestern Arkansas and in the upper Morrowan
portion of the Bird Spring Formation in southern Nevada.

Material

Lefts: 53; rights: 47.

Idiognathoides macer (Wirth, 1967)

(Fig. 44: 8)

Gnathodus macer Wirth, 1967, p. 211, Pl. 20, figs. 6-10, text Figure 11a, 11b
(*non* = c).
Idiognathoides macer (Wirth), Straka and Lane, 1970, Figures 1A, 1B.

Diagnosis

This form-species displays Class II symmetry. One-third to one-half of the anterior inner margin is flared away from the platform axis and is transversely ribbed.

Description

The blade attaches to the platform marginally. The outer margin and the posterior one-third to one-half of the inner margin are nodose. The blade is equal to or longer than the platform.

Remarks

The specimens of *Idiognathoides macer* figured by Wirth (1967, Pl. 20, figs. 6-10, text Fig. 11a, 11b) differ from the specimen illustrated herein (Fig. 44: 8) in that the former possess a wider platform upper surface and a more flared basal cavity.

Idiognathoides macer (Wirth) is distinguished from *I. sulcatus* by the flared, ribbed, inner margin of the platform. *Idiognathoides macer* is distinguished from the form-species *I. sinuatus* by possession of both lefts and rights and by a trough that continues uninterrupted to the posterior tip of the platform. Forms referred to *I.* cf. *I. macer* differ from *I. macer* in that the entire margin of the platform is ornamented with transverse ridges that continue to the bottom of the median longitudinal trough.

Occurrence

Idiognathoides macer is known to occur in the lower Namurian, Quinto Real region, West Pyrenees, Spain (Wirth, 1967). In this study, *I. macer* was recovered from the lower part of the Prairie Grove Member of the Hale Formation, Washington County, Arkansas, and *I.* cf. *I. macer* from the Primrose Member of the Golf Course Formation, Carter County, Oklahoma.

Material

Lefts: 9; rights: 12.

Type

SUI 33676.

Idiognathoides noduliferus (Ellison and Graves, 1941)

(Fig. 35: 1-15; Fig. 41: 15-17)

Cavusgnathus nodulifera Ellison and Graves, 1941, p. 4, Pl. 3, figs. 4, 6.
Streptognathodus parallelus Clarke, 1960, p. 29, Pl. V, figs. 6-8, 14, 15.

Streptognathodus japonicus Igo and Koike, 1964, p. 188, Pl. 28, figs. 5-10 (*non* figs. 11-13).

Declinognathodus nevadensis Dunn, 1966, p. 1300, Pl. 158, figs. 4a-4c, 8.

Gnathodus nodulifera (Ellison and Graves). Koike, 1967, p. 297, Pl. 3, figs. 9-12.

Idiognathoides aff. *I. nodulifera* (Ellison and Graves). Lane, 1967, p. 938, Pl. 123, figs. 9-11, 13, 17 (*non* fig. 16 = *I. sulcatus* Higgins and Bouckaert M.T.S. F).

Idiognathoides nodulifera (Ellison and Graves). Igo and Koike, 1968, p. 28, Pl. III, figs. 7-10, 12 (*non* figs. 11a, 11b = ?*Gnathodus girtyi*).

Gnathodus noduliferus (Ellison and Graves). Higgins and Bouckaert, 1968, p. 33, Pl. 2, figs. 6, 12.

Gnathodus japonicus (Igo and Koike). Higgins and Bouckaert, 1968, p. 35, Pl. 4, figs. 1, 2, 4.

[?] *Streptognathodus lateralis* Higgins and Bouckaert, 1968, p. 45, Pl. 5, figs. 1-4, 7.

Streptognathodus noduliferus (Ellison and Graves). Webster, 1969, p. 48, Pl. 4, figs. 7, 8.

Declinognathodus nodulifera (Ellison and Graves). Dunn, 1970a, p. 330, Pl. 62, figs. 1, 2 (1 = M.T.S. E; 2 = M.T.S. D).

Declinognathodus lateralis (Higgins and Bouckaert). Dunn, 1970a, p. 330, Pl. 62, figs. 5-7 (= M.T.S. E).

Declinognathodus-Neognathodus Dunn, 1970a, p. 330, Pl. 62, fig. 8 (= M.T.S. E).

Idiognathoides noduliferus (Ellison and Graves). Straka and Lane, 1970, Fig. 1A.

Diagnosis

The form-species displays Class II symmetry. The blade curves laterally from anterior axial position to meet the outer margin. No carina is present. One to three nodes, which may be fused into a ridge, are present on the anterior outer shoulder of the platform.

Description

A median longitudinal trough of variable length is developed. The upper surface is ornamented with transverse ridges or node rows that continue to the bottom of the trough.

Remarks

Of six studies on Morrowan conodonts currently in progress or recently completed, this form-species has been assigned to four different genera and four species. Although our concept of the form-species admittedly is a broad one, we feel the morphologic fluidity is best expressed in a transition series diagram, as in Straka and Lane (1970).

No evidence can be cited to suggest that the form-species in question has a direct phylogenetic relation to *Streptognathodus* Stauffer and Plummer. Furthermore, our form-generic concept of *Streptognathodus* does not include specimens with a blade that curves laterally to meet the outer margin.

The present morphologic concept of the form-genus *Gnathodus* Pander includes only those forms with a median carina that extends the full length of the platform. It is believed that *Gnathodus girtyi simplex* Dunn is the probable phylogenetic precursor of *Idiognathoides noduliferus*, but transitional specimens are lacking.

In light of the illustrations and discussion of the *Idiognathoides noduliferus-I. sulcatus* evolutionary complex by Straka and Lane (1970), it is thought that the form-species in question is best placed in *Idiognathoides*.

The relation of the occurrence of *Idiognathoides noduliferus* in the middle Morrowan and Atokan to its presence in the lower part of the Morrowan has been discussed earlier in the section on Conodont Phylogeny. Further study is necessary to clarify the three separate stratigraphic occurrences.

Occurrence

The form-species is known in the Dimple Limestone of the Marathon region of Texas (Ellison and Graves, 1941), in the Millstone Grit of the Midland Valley, Scotland (Clarke, 1960), in the lower part of the Omi Limestone, Nishikubiki County, Niigata Prefecture, central Japan (Igo and Koike, 1964), in the Morrowan of the Bird Spring Formation, Clark and Lincoln Counties, Nevada (Dunn, 1966; Webster, 1969), in the upper part of the Nagoe and lower part of the Kodani Formations, Okayama Prefecture, southwest Japan (Koike, 1967), in a brachiopodal limestone in the Sungei Lembing area near Kuantan, Malaya (Igo and Koike, 1968), and in beds ranging in age from H_1 to G_2b in the Namurian of Belgium (Higgins and Bouckaert, 1968).

In this report, *Idiognathoides noduliferus* is recorded from the Target Limestone Lentil, Lake Ardmore and Academy Church Shale Members of the Springer Formation, and the Primrose Member of the Golf Course Formation, Carter County, Oklahoma. In Washington County, Arkansas, *Idiognathoides noduliferus* is recorded from the Cane Hill and lower part of the Prairie Grove Members of the Hale Formation and in northeastern Oklahoma from the lower part of the Hale Formation.

Material

The number of specimens recovered representing each Morphologic Transition Series (Straka and Lane, 1970) within *Idiognathoides noduliferus* are as follows.

M.T.S.	left	right
B	0	91
C	73	0
A	17	2
D	25	15
E	34	21
I. noduliferus	110	90
Total	259	219

Types

SUI 33611-33615.

Idiognathoides sinuatus Harris and Hollingsworth, 1933

(Fig. 37: 14, 15, 18, 20, 23-26, 36; Fig. 41: 1-14, 20-27)

Idiognathoides sinuata Harris and Hollingsworth, 1933, p. 201, Pl. 1, fig. 14; Lane, 1967, p. 937, Pl. 119, figs. 1-9, 12-15, Pl. 123, figs. 7, 8, 12 (fig. 8 = holotype); Higgins and Bouckaert, 1968, p. 40, Pl. 2, fig. 14, Pl. 4, figs. 5, 8, 9, Pl. 5, fig. 11.

Idiognathodus corrugata Harris and Hollingsworth, 1933, p. 202, Pl. 1, figs. 7, 8a, 8b (figs. 8a, 8b = lectotype selected herein).

Polygnathodella ouachitensis Harlton, 1933, p. 15, Pl. 4, figs. 14a-14c; Koike, 1967, p. 309, Pl. 3, figs. 3-5.

Cavusgnathus sinuata (Harris and Hollingsworth). Ellison and Graves, 1941, p. 5, Pl. 3, figs. 1, 5, 7.

Polygnathodella attenuata (Harris and Hollingsworth). Ellison and Graves, 1941, p. 8, Pl. 3, figs. 11, 13 (*non* figs. 14?, 15 = *I. attenuatus*).

Gnathodus opimus Igo and Koike, 1964, p. 189, Pl. 28, fig. 18 (*non* figs. 15-17 = *I. sulcatus*); Webster, 1969, p. 33, Pl. 5, figs. 20, 21 (*non* fig. 19 = ?*I. sulcatus*).

Idiognathoides corrugata (Harris and Hollingsworth). Lane, 1967, p. 939, Pl. 122, figs. 1, 2, 4-7, 9-11; Higgins and Bouckaert, 1968, p. 39, Pl. 5, fig. 9.

Idiognathoides convexa (Ellison and Graves). Higgins and Bouckaert, 1968, p. 39, Pl. 4, fig. 3.

Idiognathoides sinuatus Harris and Hollingsworth. Dunn, 1970a, p. 335, Pl. 63, figs. 14, 15, 22, 23 (*non* fig. 21 = *I. sulcatus*).

Idiognathoides corrugatus (Harris and Hollingsworth). Dunn, 1970a, p. 335, Pl. 63, figs. 16-18, 25.

Diagnosis

The species displays Class IIIb symmetry. The blade attaches to the left margin in the left element, and a median longitudinal trough of variable length is present on the upper surface of the platform. No distinct carina is developed, and the left (outer) margin of the transversely ridged platform is elevated above the right margin. The free blade attaches to the right margin of the platform in the right element, and a short median longitudinal trough is present in the anterior third of the platform. No carina is developed. The margins of the transversely ridged platform are of equal height.

Description

Left element: The left element of *Idiognathoides sinuatus* displays features characteristic of the holotype of the form-sepcies *I. sinuatus* Harris and Hollingsworth.

Right element: The right conodont element of *Idiognathoides sinuatus* displays features characteristic of the lectotype of the form-species *Idiognathoides corrugatus* (Harris and Hollingsworth).

Remarks

In addition to the description of the form-species *Idiognathoides sinuatus* and *I. corrugatus* (see Lane, 1967, p. 937, 939), the following additional morphologic characters were noted in collections studied for the current report.

In some left elements, the left third of the platform is represented by an elevated but level, transversely corrugated ridge, from which the corrugated upper platform surface drops abruptly into the median trough or declines sharply to the inner margin. A node on the anterior outer side of the platform may be present (Fig. 41: 25). The anterior inner margin may be slightly to moderately flared and ornamented with transverse ridges that tend to radiate from the median trough to the inner margin (Fig. 41: 27). Upper Morrowan forms seem to have a more strongly flared basal cavity (Fig. 41: 20, 22) that dips moderately away from the ornamented upper surface of the platform, rather than dropping off vertically to meet the cavity flare as in older Morrowan forms. Also, upper Morrowan representatives possess a narrower, ornamented platform surface (Fig. 41: 20, 22).

There is a tendency in some right elements toward development of a broad, ornamented upper surface of the platform (Fig. 41: 3, 14), a slight flaring of the transversely ridged anterior inner margin (Fig. 41: 3, 7, 8), and a node or circular ridge on the outer anterior margin (Fig. 41: 3, 6-8). Upper Morrowan right elements (Fig. 41: 4), as in the left elements, develop a more strongly flared basal cavity that dips moderately away from the ornamented upper surface of the platform, rather than dropping off vertically to meet the cavity flare.

The reasons for the synonymy of the form-species *Idiognathoides sinuatus* Harris and Hollingsworth and *Idiognathoides corrugatus* (Harris and Hollingsworth) are the same as those discussed in Remarks for *Adetognathus lautus* (Gunnell). Because most previous authors did not present range and numerical occurrence data for the two form-species, our synonymy must be used with some caution.

We believe that *Idiognathoides sinuatus* originated phylogenetically from the symmetrically paired *I. noduliferus–I. sulcatus* morphologic complex (Straka and Lane, 1970).

Occurrence

Idiognathoides sinuatus is known to occur in the Union Valley Formation NE1/4 of sec. 29, T. 3 N., R. 7 E. (not "NW 1/4. . ." as stated by Harris and Hollingsworth, 1933, p. 202), Pontotoc County, Oklahoma. *Idiognathoides sinuatus* ranges from the Prairie Grove Member of the Hale Formation into the base of the Dye Shale Member of the Bloyd Formation in Washington County, Arkansas. It is known to occur from the Hale Formation to the middle of the Bloyd Formation in northeastern Oklahoma and throughout the Primrose Member of the Golf Course Formation, Carter County, Oklahoma.

Material

Lefts: 1,106; rights: 1,307.

Types

SUI 33643-33647, 33740-33753.

Idiognathoides sulcatus Higgins and Bouckaert, 1968

(Fig. 36: 1-16, 18-20, 23, 24; Fig. 39: 1-10)

Gnathodus opimus Igo and Koike, 1964, p. 189, Pl. 28, figs. 15-17 (*non* fig. 18 = form-species *I. sinuatus*); Igo and Koike, 1965, p. 89, Pl. 9, figs. 1-4 (*non* figs. 5-8); Koike, 1967, p. 299, Pl. 1, figs. 20a-20b, 21; Webster, 1969, p. 33, Pl. 5, fig. 20 (*non* figs. 19, 21 = *I. sinuatus*).

Idiognathoides aff. *I. nodulifera* (Ellison and Graves). Lane, 1967, p. 938, Pl. 123, fig. 16 (*non* figs. 9-11, 13, 17 = *I. noduliferus*); ?Palmieri, 1969, p. 7, Pl. 3, figs. 10, 11 (*non* Pl. 3, figs. 9, 16-17, 24-25, Pl. 4, figs. 10-13, 26 = *Idiognathoides* spp.).

Idiognathoides sp. A. Lane, 1967, p. 938, Pl. 123, figs. 14, 15, 18, 19.

Idiognathoides cf. *I. sinuatus* Harris and Hollingsworth. Palmieri, 1969, p. 7, Pl. 4, figs. 6-9 (*non* Pl. 3, figs. 1-4 = *Idiognathoides* sp.).

Idiognathoides sinuatus Harris and Hollingsworth. Dunn, 1970a, p. 335, Pl. 63, fig. 21 (*non* figs. 14, 15, 22, 23 = *I. sinuatus*).

Idiognathoides opimus (Igo and Koike). Dunn, 1970a, p. 335, Pl. 63, figs. 24, 28-30.

Idiognathoides sulcata Higgins and Bouckaert, 1968, p. 41, Pl. 4, figs. 6, 7, Pl. 6, figs. 1-6.

Idiognathoides sulcatus Higgins and Bouckaert. Straka and Lane, 1970, p. 42, Fig. 1A.

Diagnosis

The form-species displays Class II symmetry. The blade-outer margin outline is straight or gently bowed outwardly. The parallel margins of the platform vary from being nodose to ridged and are of equal height.

Description

The median longitudinal trough continues uninterrupted to the posterior tip. The gnathodid-type basal cavity is elliptical in upper view, and its long axis is oriented at about 45° to the median axis of the platform. The basal cavity flare on the inner side is anterior to that of the outer.

Remarks

Lane (1967, p. 938-939) stated that this form-species is invariably right-sided. Additional collections now show that such a statement is incorrect because left-sided elements have been recovered. Forms that Igo and Koike (1964, Pl. 28, figs. 15-17; 1965, Pl. 9, figs. 1-4) and Koike (1967, Pl. 1, figs. 20a, 20b, 21) referred to *Gnathodus opimus* Igo and Koike are considered to be conspecific with *Idiognathoides sulcatus*. However, the holotype of *G. opimus* appears to be conspecific with the form-species *I. sinuatus* Harris and Hollingsworth.

Idiognathoides sulcatus is distinguished from *I. noduliferus* by the absence of a node on the outer anterior margin and from *I. sinuatus* by the absence of incomplete development of transverse-ridged ornamentation on the platform.

Occurrence

Idiognathoides sulcatus is known to occur in the lower part of the Omi Limestone in Nishikubiki County, Niigata Prefecture, central Japan (Igo and Koike, 1964, Pl. 28, figs. 15-17); in the Akiyoshi Limestone at Yobara, Ofuku, Yamaguchi Prefecture, southwestern Japan (Igo and Koike, 1965, Pl. 9, figs. 1-4); in the Kodani Formation, Okayama Prefecture, southwestern Japan (Koike, 1967); in the Bird Spring Formation of Clark and Lincoln Counties, Nevada (Webster, 1969); in the Kinderscoutian (R1), Marsdenian (R2), and Lower Westphalian (G2) of the type Namurian area, Belgium (Higgins and Bouckaert, 1968); and in limestones in close proximity to the Wondai Series, Murgon, Queensland, Australia (Palmieri, 1969). In this study, the species ranges from the Academy Church Shale Member of the Springer Formation to near the top of the Primrose Member of the Golf Course Formation, Carter County, Oklahoma, and occurs in the Prairie Grove Member of the Hale Formation and the Brentwood Limestone Member of the Bloyd Formation in northwestern Arkansas and in the Hale and Bloyd Formations in northeastern Oklahoma.

Material

The number of specimens representing each Morphologic Transition Series within *Idiognathoides sulcatus* are as follows.

M.T.S.	left	right
F	0	129
G	15	0
A	15	3
H	65	54
I	34	83
I. sulcatus	491	470
Total	620	739

Types

SUI 33634-33640, 33725-33730.

Idiognathoides sulcatus sulcatus Higgins and Bouckaert, 1968

(Fig. 36: 1-6, 18-20, 23, 24)

Idiognathoides sulcata Higgins and Bouckaert, 1968, p. 41, Pl. 4, figs. 6, 7.
Idiognathoides sulcatus Higgins and Bouckaert. Straka and Lane, 1970, Fig. 1A
    (*non* morphotypes F5, G5, A2, 3, H4, I5).

Diagnosis

The parallel platform margins are unornamented or ridged, with the ridges ending near the center of the longitudinal trough. The blade is the same length or slightly longer than the platform. The platform in lateral view is arched downward from the blade-platform contact.

Remarks

The morphologies constituting our form-subspecific concept are illustrated by Straka and Lane (1970, Fig. 1A, morphotypes F1-4, G1-3, H1-3, I1-4). *Idiognathoides sulcatus sulcatus* is distinguished from the left element of *I. sinuatus* by possessing platform margins of equal height and by the lack of transverse ridges on the platform. *Idiognathoides macer* is separated from *I. sulcatus sulcatus* by the outward flare of the anterior inner platform margin in the former.

Occurrence

Idiognathoides sulcatus sulcatus is known to occur from the Kinderscoutian (R1) into the Lower Westphalian (G2b) of the type Namurian area of Belgium (Higgins and Bouckaert, 1968). In this study, the form-subspecies occurs in the Academy Church Shale Member of the Springer Formation and the Primrose Member of the Golf Course Formation in Carter County, Oklahoma; in the Prairie Grove Member of the Hale Formation in northwestern Arkansas; and in the Hale Formation in northeastern Oklahoma.

Material

Lefts: 553; rights: 693.

Types

SUI 33634-33640.

Idiognathoides sulcatus parvus Higgins and Bouckaert, 1968

(Fig. 39: 1-10)

Idiognathoides sp. A. Lane, 1967, p. 938, Pl. 123, figs. 14, 15, 18, 19.
Idiognathoides sulcatus parvus Higgins and Bouckaert, 1968, p. 41, Pl. 6, figs.
    1-6.

Diagnosis

In lateral view, the blade is approximately one-and-a-half times longer than the platform. The upper profile of the platform is flat but is flexed slightly downward at the blade-platform junction. The margins of the platform consist of a nodose ridge.

Remarks

Idiognathoides sulcatus parvus is distinguished from the nominate form-subspecies by the high flat platform that is flexed slightly downward at the blade-platform junction. Also, the margins of the platform in *I. sulcatus parvus* are nodose rather than transversely ornamented ridges as in *I. sulcatus sulcatus*. Furthermore, the blade in *I. sulcatus parvus* is approximately one-and-a-half times longer than the platform, and the posterior 4 to 6 denticles of the blade are flat and fused into a ridge.

Occurrence

The form-subspecies occurs in the Lower Westphalian (G2b) in the type Namurian region, Belgium (Higgins and Bouckaert, 1968). Herein the subspecies is recorded from the upper part of the Brentwood Member of the Bloyd Formation in Washington County, Arkansas, the lower part of the Bloyd Formation in northeastern Oklahoma, and the Primrose Member of the Golf Course Formation in Carter County, Oklahoma.

Material

Lefts: 60; rights: 50.

Types

SUI 33725–33730.

Form-genus *Neognathodus* Dunn, 1970a

Type-species: *Polygnathus bassleri* Harris and Hollingsworth, 1933, p. 198, Pl. 1, figs. 13a–13e (figs. 13d, 13e = lectotype selected and refigured by Lane, 1967, p. 935, Pl. 123, figs. 2, 4).

Diagnosis

The platform form-genus displays Class II symmetry. In upper view, the carina extends to or near the posterior tip. The long free blade meets the platform centrally or subcentrally. The outer margin of the platform may be reduced or absent. Deep adcarinal grooves are developed. In lower view, a large, deep, asymmetrical basal cavity is present at the posterior end of the platform.

Remarks

The genus *Neognathodus* originated phylogenetically from the *Idiognathoides noduliferus-I. sulcatus* evolutionary complex (Straka and Lane, 1970). The deep adcarinal grooves distinguish early forms of *Neognathodus* from *Gnathodus* and give *Neognathodus* a polygnathid appearance in upper view. Furthermore, the carina in *Gnathodus* extends to the posterior tip of the platform; whereas, in some representatives of *Neognathodus*, the carina may end just before reaching the posterior tip. *Neognathodus* is distinguished from *Idiognathoides* by possession of a median carina that extends to or nearly to the posterior tip.

Neognathodus is a homeomorph of *Gnathodus*. *Gnathodus* originated from *Spathognathodus* and *Neognathodus* from *Idiognathoides*.

Bicarinodus Jones, 1941, may represent the senior synonym of *Neognathodus*, but the type-species (*B. oherni*) is lost and the illustrations and descriptions are inadequate for identification. Consequently, *Bicarinodus* Jones is considered to be a *nomen dubium*.

Range

Neognathodus appears in the lower Morrowan (base of *Neognathodus bassleri symmetricus* Zone), based on this report, and ranges to the top of the Desmoinesian (Ellison, 1941).

Neognathodus bassleri (Harris and Hollingsworth, 1933)

(Fig. 37: 16, 17, 19, 22, 31, 32, 37-39; Fig. 39: 16-18, 21-24; Fig. 42: 17-24)

Polygnathus bassleri Harris and Hollingsworth, 1933, p. 198, Pl. 1, figs. 13a-13e.
Polygnathus wapanuckensis Harlton, 1933, p. 15, Pl. 4, figs. 13a-13c.
Gnathodus wapanuckensis (Harlton). Ellison and Graves, 1941, Pl. 2, figs. 13-17;
    Koike, 1967, p. 300, Pl. 1, figs. 22-25.
Streptognathodus wapanuckensis (Harlton). Elias, 1956a, p. 120, Pl. 3, figs. 67-69;
    ?Wirth, 1967, p. 236, Taf. 20, figs. 11, 13.
Gnathodus bassleri (Harris and Hollingsworth). Lane, 1967, p. 934, Pl. 120, figs.
    1-5, 9-15, 17, Pl. 121, figs. 6, 9, Pl. 123, figs. 1-6 (lectotype = figs. 2, 4);
    Lane and Straka *in* Lane and others, 1971, Pl. 1, figs. 7-10.
[*non*] *Gnathodus bassleri* (Harris and Hollingsworth). Webster, 1969, p. 29, Pl.
    5, figs. 9, 14, 15 [=*N. colombiensis* (Stibane)]; Webster *in* Lane and others,
    1971, Pl. 1, fig. 26 [=*N. colombiensis* (Stibane)]; Merrill *in* Lane and others,
    1971, Pl. 1, figs. 29, 30 [= *N. colombiensis* (Stibane)].
Neognathodus bassleri (Harris and Hollingsworth). Dunn, 1970a, p. 336, Pl. 64,
    figs. 1, 12, 13 [*non* fig. 14 = *N. colombiensis* (Stibane)].

Remarks

Diagnosis, description, and discussion of this form-species have been given by Lane (1967, p. 934-935).

Neognathodus bassleri bassleri (Harris and Hollingsworth, 1933)

(Fig. 37: 16, 17, 19; Fig. 42: 17-24)

Polygnathus bassleri Harris and Hollingsworth, 1933, p. 198, Pl. 1, figs. 13a-13c.
Polygnathus wapanuckensis Harlton, 1933, p. 15, Pl. 4, figs. 13a-13c.
Streptognathodus wapanuckensis (Harlton). Elias, 1956a, p. 120, Pl. III, figs. 67-69.
Gnathodus bassleri bassleri (Harris and Hollingsworth). Lane, 1967, p. 935, Pl. 120, figs. 1, 3-5, 9-12, 15, Pl. 123, figs. 1-6; Lane and Straka *in* Lane and others, 1971, Pl. 1, figs. 9, 10.
[*non*] *Gnathodus bassleri bassleri* (Harris and Hollingsworth). Merrill *in* Lane and others, 1971, Pl. 1, fig. 30 [= *N. colombiensis* (Stibane)].
Neognathodus bassleri (Harris and Hollingsworth). Dunn, 1970a, p. 336, Pl. 64, figs. 1, 13; [*non* figs. 12, 14; 12 = *N. bassleri symmetricus* (Lane) and 14 = *N. colombiensis* (Stibane)].

Remarks

Neognathodus bassleri bassleri originated from *N. bassleri symmetricus* (Lane). Specimens of *N. bassleri symmetricus* that morphologically approach but do not duplicate *N. bassleri bassleri* have been found within the range of the former. However, no specimens of *N. bassleri symmetricus* have been found within the range of *N. bassleri bassleri*. In *N. bassleri bassleri*, the inner margin is more strongly flared and the carina is closer to the outer margin than in *N. bassleri symmetricus*. Also, the inner margin of *N. bassleri bassleri* is elevated above the outer margin. The fact that the two form-subspecies do not overlap vertically in Arkansas and Oklahoma constitutes one of the most distinctive biostratigraphic subdivisions in the Morrowan.

Occurrence

In addition to occurrences cited by Lane (1967, p. 935), *Neognathodus bassleri bassleri* was found in the Woolsey Member of the Bloyd Formation, Washington County, Arkansas, and in the lower part of the Bloyd Formation of northeastern Oklahoma. It is also found in the middle and upper part of the Primrose Member of the Golf Course Formation of Carter County, Oklahoma.

Material

Lefts: 424; rights: 487.

Types

SUI 33655, 33768-33773.

Neognathodus bassleri symmetricus (Lane, 1967)

(Fig. 37: 22, 31, 32, 37–39; Fig. 39: 16–18, 21–24)

Gnathodus wapanuckensis (Harlton). Ellison and Graves, 1941, Pl. 2, figs. 13–17;
    Koike, 1967, p. 300, Pl. 1, figs. 22, 24 (figs. 23, 25a, 25b = ?).
Gnathodus bassleri symmetricus Lane, 1967, p. 935, Pl. 120, figs. 2, 13, 14, 17,
    Pl. 121, figs. 6, 9.
[non] Gnathodus bassleri symmetricus Lane. Merrill in Lane and others, 1971,
    Pl. 1, fig. 29 [= N. colombiensis (Stibane)].
Neognathodus bassleri (Harris and Hollingsworth). Dunn, 1970a, p. 336, Pl. 64,
    fig. 12.

Remarks

The phylogenetic origin of Neognathodus bassleri symmetricus is discussed by Straka and Lane (1970). See Lane (1967, p. 935) for the diagnosis of this form-subspecies.

Occurrence

In addition to those occurrences cited by Lane (1967, p. 936), N. bassleri symmetricus is known to occur in the lower part of the Kodani Formation, Okayama Prefecture, southwest Japan (Koike, 1967, p. 300). The form-subspecies was found in the Hale Formation and lower part of the Bloyd Formation in northeastern Oklahoma and first appears near the base of the Primrose Member of the Golf Course Formation, Carter County, Oklahoma.

Material

Lefts: 477; rights: 438.

Types

SUI 33651–33654, 33733–33736.

Form-genus *Rhachistognathus* Dunn, 1966

Type-species: Rhachistognathus primus Dunn, 1966, p. 1301, Pl. 157, figs. 1a–1c.

Diagnosis

The members of the form-genus display Class IIIa symmetry. For further form-generic characteristics, the reader is referred to Dunn (1966, p. 1301).

Remarks

Rhachistognathus is practically indistinguishable from the Lower Devonian form-genus *Eognathodus* Philip, 1965, and can be morphologically grouped with other double-rowed spathognathodids known from the Middle Devonian, Upper Devonian, Kinderhook (Lower Mississippian) (the latter two occurrences referred to *Bispathodus* by Müller, 1962), Meramec (Upper Mississippian) and Wolfcamp (lower Permian). Therefore, our usage of this form-genus is of a provisional nature and awaiting a better understanding of the relation between the separate occurrences of the double-rowed spathognathodids.

Range

Rhachistognathus appears at the base of the uppermost Mississippian *Rhachistognathus muricatus* Zone and ranges at least to the top of the *Neognathodus bassleri bassleri* Zone in southern Nevada.

Rhachistognathus muricatus (Dunn, 1965)

(Fig. 35: 16, 17, 24, 30, 31)

Cavusgnathus muricata Dunn, 1965, p. 1147, Pl. 140, figs. 1, 4.
[?] *Idiognathoides minuta* Higgins and Bouckaert, 1968, p. 40, Pl. 6, figs. 7–12.
Gnathodus muricatus (Dunn). Webster, 1969, p. 32, Pl. 5, figs. 1, 2, 4–7 (*non* fig. 3 = *R. primus*).
Rhachistognathus muricatus (Dunn). Dunn, 1970a, p. 338, Pl. 61, figs. 5–7.
Spathognathodus muricatus (Dunn). Lane and Straka *in* Lane and others, 1971, Pl. 1, fig. 1.

Diagnosis

This form-species displays Class IIIa symmetry. The blade is always attached to the left margin of the platform and both lefts and rights are present. The platform is flanked by two parallel rows of circular to slightly oval nodes; this produces a slitlike trough and usually continues to the posterior end of the platform. In some specimens the left row of nodes ends about two-thirds of the way posteriorly, and an offset but centrally located row of nodes continues to the posterior end of the platform.

Remarks

Rhachistognathus muricatus appears in the uppermost Mississippian at the base of the *R. muricatus* Zone and ranges at least to the top of the *N. bassleri bassleri* Zone in southern Nevada. Within the Lower Pennsylvanian *Rhachistognathus primus* Zone, the form-species intergrades with *R. primus*.

Rhachistognathus muricatus may be distinguished from form-species of *Cavusgnathus* principally on the basis of symmetry characteristics and by the nodose nature of the platform margins. Species of the latter genus are characterized by Class IV symmetry, as the individual conodont elements are right-sided with indistinct curvature.

Occurrence

Rhachistognathus muricatus has been reported from the upper part of the Indian Springs and the lower part of the Bird Spring Formations in southern Nevada (Dunn, 1965, 1966, 1970a; Webster, 1969; Lane and others, 1972). The species is also known to occur in the La Tuna Limestone of west Texas (Lane and others, 1972). In southern Oklahoma, *R. muricatus* occurs in the Target Limestone Lentil of the Springer Formation at Locality 5, and a single specimen from the middle portion of the Primrose Member of the Golf Course Formation at Locality 3 may be reworked. Specimens occurring at the base of the Hale Formation in northeastern Oklahoma (Localities 16 and M-25) may be reworked because of their broken and worn condition within an otherwise well-preserved fauna. Occurrence of one poorly preserved specimen of *Rhachistognathus muricatus* in the top of the Pitkin Limestone at Locality M-5 is considered as probably originating from the conglomerate that separates Pitkin from Hale strata.

Material

Lefts: 73; rights: 47; 7 with indeterminate curvature.

Types

SUI 33605, 33607.

Rhachistognathus primus Dunn, 1966

(Fig. 35: 18-23, 25-29, 32-40; Fig. 44: 6)

Rhachistognathus prima Dunn, 1966, p. 1301, Pl. 157, figs. 1, 2.
Cavusgnathus transitoria Dunn, 1966, p. 1299, Pl. 157, figs. 9, 13.
Gnathodus muricatus (Dunn). Webster, 1969, p. 32, Pl. 5, fig. 3 (*non* figs. 1, 2, 4-7 = *R. muricatus*).
Rhachistognathus primus Dunn. Dunn, 1970a, p. 338, Pl. 63, figs. 26, 27.
Rhachistognathus transitorius (Dunn). Dunn, 1970a, p. 339, Pl. 61, fig. 14.
Spathognathodus muricatus (Dunn). Lane and Straka *in* Lane and others, 1971, Pl. 1, fig. 2.

Diagnosis

The form-species displays Class IIIa symmetry. The conodont elements are all left-sided with respect to blade attachment. A nodose carina is strongly developed in the posterior part of the trough and either dies out as a series of nodes alongside the posterior end of the blade or is continuous with the blade as a ridge. The upper surface is characterized by development of irregular sized and spaced nodes, or ridges along the margins, or both.

Description

A median trough may or may not be developed on the platform. The right margin projects farther to the anterior than does the left margin. In some cases curvature is not present, and a nearly symmetrical unit is approached.

In lateral view, the free blade is long and rises in upper outline from the blade-platform junction to its apex near the anterior end. A characteristic, nondenticulated straight profile is developed in the upper outline of the blade posterior to the blade-platform junction and extends posteriorly one-quarter to one-third the length of the platform.

Remarks

Transitional specimens in our collection demonstrate that within the *Rhachistognathus primus* Zone, *R. muricatus* and *R. primus* are totally intergradational, via the *R. transitorius* morphotype. The specimens figured herein (Fig. 35: 30, 18) represent these end-members, respectively, and other figured specimens (Fig. 35: 29, 23) illustrate transitional morphologic variation (*R. transitorius* morphotype) between the two form-species. The change in morphology from the end-member (Fig. 35: 30) to the other (Fig. 35: 18) occurs along two developmental lines: (1) The blade-left margin of Figure 35: 30 is continuous and straight, and a carina is weakly developed in the trough near the posterior end of the platform. The carina projects anteriorly one-third the length of the platform and is juxtaposed with the nodose left margin of the platform at its anterior termination. Morphologic alteration from Figure 35: 30 to 29 to 23 to 18 includes gradual right-migration of the anterior one-third of the left margin and increase in size of the carinal nodes. Carinal development also includes anterior extension of the carina in the trough. Progressive margin migration and carinal development result in eventual fusion of the migrated margin and carina into a prominent, sharp, straight, median ridge or carina that extends the entire length of the platform (Fig. 35: 18). (2) A second morphologic change through the figured specimens from end-member (Fig. 35: 30) to end-member (Fig. 35: 18) results in a gradual loss of marginal ornamentation. As the left anterior margin begins to migrate and the ridgelike carina develops, there is a progressive loss of nodes from both margins, which proceeds anteriorly from the posterior tip.

It is notable that the morphologic alteration also seems to involve a symmetry transition from curvature pairs of Figure 35: 30-type to indistinct curvature typifying specimens like Figure 35: 18. Those element-pairs of the morphologic end-member represented by Figure 35: 18 result in no change of symmetry class characteristics. Such pairs may still be appropriately categorized as Class IIIa, based upon recognition of their relation to curvature pairs through morphologic transition as discussed above. However, if viewed by themselves, such elements with indistinct curvature can be appropriately designated as hypothetical Class IV symmetry.

The morphologic intergradation between *R. muricatus* and *R. primus* via the

R. transitorius morphotype is restricted to the *R. primus* Zone. *Rhachistognathus primus* (a morphologic concept that includes *R. transitorus*) is restricted in range to the *R. primus* Zone; whereas in Nevada, *R. muricatus* ranges as low as the Upper Mississippian *R. muricatus* Zone and as high as the middle Morrowan *N. bassleri bassleri* Zone.

We view *Rhachistognathus muricatus* and *R. primus* as belonging to the double-rowed group of spathognathodids. Those features that demonstrate this association include the slitlike trough, blade attachment to the left margin, Class IIIa symmetry, and the basal cavity, all of which are alike in *Rhachistognathus* and most other representatives of the double-rowed spathognathodids. In addition to the above features, the presence of a poorly developed carina typifies *R. primus* and *R. muricatus*. Klapper and Ormiston (1969, p. 23) noted in *S. sulcatus* from the Lower Devonian that "a third row, represented by a thin medial ridge is present in the posterior fourth of some specimens." They also noted a "remarkable flat" portion in the lateral view of the upper profile of *S. sulcatus*. As previously stated (see Diagnosis and Description) these features are also present in *R. muricatus* and *R. primus*. In light of these morphologic similarities, we consider these form-species as members of the double-rowed spathognathodid group, and particularly as homeomorphs of *Spathognathodus sulcatus* (Philip).

Occurrence

Rhachistognathus primus has been reported from the lower Morrowan of the Bird Spring Formation, Clark County, Nevada (Dunn, 1966,1970a; Webster, 1969).

In southern Oklahoma a large fauna including *R. primus* was recovered from the Target Limestone Lentil of the Springer Formation at Locality 5.

In northwestern Arkansas, the form-species occurs in the Cane Hill Member of the Hale Formation at Localities 17-19. Occurrences of the form-species at Localities 16 and 20 may be due to reworking. Dunn (1966) reported the species from the Magdalena Limestone of west Texas.

Material

Lefts: 15; rights: 21; 12 left-sided elements, with indeterminate curvature.

Types

SUI 33601-33604, 33606, 33608-33610, 33795.

Form-genus *Spathognathodus* Branson and Mehl, 1941a

Spathognathodus Branson and Mehl, 1941a, p. 98 (*pro Spathodus* Branson and Mehl).

Type-species: *Spathodus primus* Branson and Mehl, 1933, p. 46 (original designation).

Spathognathodus minutus (Ellison, 1941)

(Fig. 44: 7, 12)

Spathodus minutus Ellison, 1941, p. 120, Pl. 20, figs. 50–52.
Spathognathodus minutus (Ellison). Ellison and Graves, 1941, Pl. 2, figs. 1, 3, 5; Youngquist and Downs, 1949, p. 169, Pl. 30, fig. 4; Sturgeon and Youngquist, 1949, p. 385, Pl. 74, figs. 9–11, Pl. 75, fig. 19; Rexroad and Burton, 1961, p. 1156, Pl. 141, figs. 10, 11; Murray and Chronic, 1965, p. 606, Pl. 72, figs. 29, 30; Igo and Koike, 1965, p. 88, Pl. 9, figs. 16–18; Koike, 1967, p. 311, Pl. 3, figs. 39–42; Webster, 1969, p. 44, Pl. 7, fig. 4; Dunn, 1970a, p. 339, Pl. 61, figs. 27, 30.

Occurrence

Spathognathodus minutus occurs in the Prairie Grove Member of the Hale Formation and the Brentwood Limestone Member of the Bloyd Formation in northwestern Arkansas and in the Hale and Bloyd Formations of northeastern Oklahoma.

Material

30.

Type

SUI 33796.

Spathognathodus n. sp.

(Fig. 44: 15, 16)

Spathognathodus n. sp. Lane, 1967, p. 939, Pl. 120, figs. 6–8.

Remarks

This form-species is left unnamed due to the lack of adequate material.

Occurrence

The form-species is known to occur in the upper part of the Prairie Grove Member of the Hale Formation and the Brentwood Limestone and Dye Shale (caprock of the Baldwin coal) Members of the Bloyd Formation in Washington County, Arkansas. *Spathognathodus* n. sp. also occurs in the upper Hale and lower Bloyd Formations of northeastern Oklahoma.

Material

17.

Types

SUI 33802.

Form-genus *Streptognathodus* Stauffer and Plummer, 1932

Type-species: *Streptognathodus excelsus,* Stauffer and Plummer, 1932, p. 48, Pl. 4, figs. 2, 5.

Remarks

Streptognathodus is distinguished from *Idiognathodus* by possession of a median longitudinal trough that interrupts the posterior transverse ridges. The median trough bears an anterior carina extending about one-third the length of the platform.

Range

Streptognathodus first occurs in the Woolsey Member of the Bloyd Formation (Morrowan) in this study and is known to range into the Permian (Hass, 1962, p. W62).

Streptognathodus expansus Igo and Koike, 1964

(Fig. 43: 9, 16–18, 21–26)

Streptognathodus expansus Igo and Koike, 1964, p. 189, Pl. 28, fig. 14; Koike, 1967, p. 312, Pl. 3, figs. 6, 8, 16 (figs. 7, 17 = ?); Webster, 1969, p. 47, Pl. 6, figs. 1–5; Dunn, 1970a, p. 339, Pl. 62, figs. 18–20.

Diagnosis

The form-species displays Class IV symmetry.

Description

Left element: Length of the median free blade approximately equals that of the platform. A slitlike median trough generally extends from the posterior end of the carina to near the posterior tip of the platform. However, in some cases this slitlike trough is present only in the posterior one-third of the platform. The carina may be flanked by 2 rostral ridges and by a small inner lobe ornamented with 1 to 5 nodes. The upper surface is ornamented with well-developed transverse ridges that end at the slitlike trough and curve anteriorly near the margins.

In lateral view, the unit is arched with the highest point at the junction of the blade and carina.

In lower view, a gnathodid-type of cavity is developed. The cavity continues on the lower side of the free blade as a slit.

Right element: No right-sided element is known to exist with certainty. However, on an empirical basis, Dunn (1970a) suggested that *S. expansus* should be associated with the form-species *S. suberectus* Dunn.

Remarks

Koike (1967, Pl. 3, figs. 7, 17) figured several right-sided specimens, which he referred to *Streptognathodus expansus.* One figured specimen (7a, 7b) has a peculiar V-shaped upper outline in lateral view, and another (17a, 17b) possesses a very narrow platform. These two figured specimens are herein excluded from *S. expansus.*

The range of variation of *Streptognathodus expansus* includes forms that closely resemble *Idiognathodus.* The median trough, a diagnostic feature of *Streptognathodus*, tends to be reduced in some forms (Fig. 43: 24, 25). *Streptognathodus expansus* may have been a substitute left element in the element-pair species *I. sinuosis* Ellison and Graves or paired with *S. suberectus* as suggested by Dunn (1970a, p. 339–340). However, the rarity of *S. expansus* and *S. suberectus* precludes a definite association at this time. Until a definite association is established, *S. expansus* will remain in Class IV symmetry.

Occurrence

Streptognathodus expansus is known to occur in the Omi Limestone, Nishikubiki County, Niigata Prefecture, southwestern Japan; in the Bird Spring Formation in Lincoln and Clark Counties, Nevada; and in the middle to upper part of the Bloyd Formation, northeastern Oklahoma.

Material

Lefts: 12; rights: 0.

Types

SUI 33787–33791.

Streptognathodus spp.

(Fig. 37:33-35; Fig. 40:19-21)

Remarks

Although specimens displaying the diagnostic morphologic features of *Streptognathodus* are placed here, their taxonomic status is puzzling. Lefts and rights included here seem to intergrade with the lefts and rights of the element-pair species *Idiognathodus sinuosis* Dunn. However, too few specimens were recovered to demonstrate adequately such an intergradation.

Material

Lefts: 2; rights: 5.

Types

SUI 33737-33739.

FIGURES 32–44

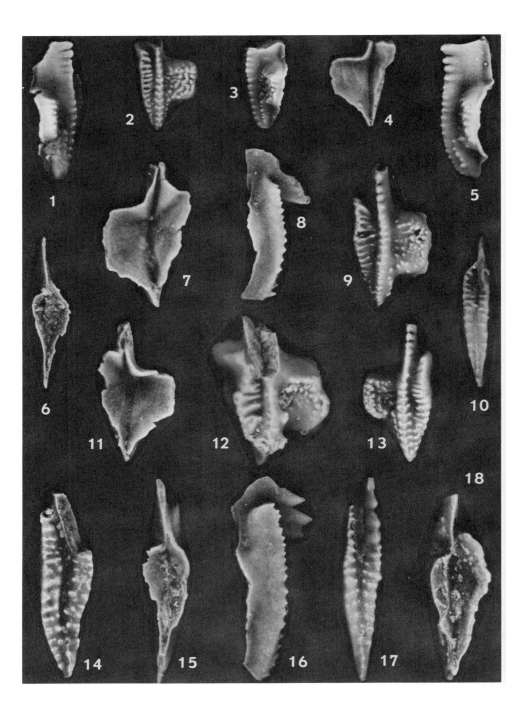

FIGURE 32. *GNATHODUS* AND *CAVUSGNATHUS*

All figured specimens are from the Goddard Formation. All illustrations are unretouched photographs and are × 31.

1-5, 7, 9, 11-13—*Gnathodus bilineatus* (Roundy): 1, 11, 13, outer lateral, lower, and upper views of *morphotype* δ, SUI 33661 (7-2); 2-4, upper, outer lateral, and lower views of *morphotype* β, SUI 33662 (11-33); 5, 7, 9, outer lateral, lower, and upper views of *morphotype* α, SUI 33663 (7-2); 12, upper view of *morphotype* γ, SUI 33664 (7-7).

6, 8, 10, 14-18—*Cavusgnathus naviculus* (Hinde): 6, 8, 10, lower, lateral, and upper views of SUI 33617 (13-1); 14, 18, upper and lower views of SUI 33618 (7-1); 15-17, lower, lateral, and upper views of SUI 33616 (9-26).

FIGURE 33. *GNATHODUS* AND *ADETOGNATHUS*

Figured specimens 1-10 and 24, 26, and 27 are from the "B" Shale Member of the Springer Formation; all others are from the Goddard Formation. All illustrations are unretouched photographs and are × 31.

1-10, 24, 26, 27—*Gnathodus girtyi intermedius* Globensky: 1, 2, outer lateral and upper views of SUI 33656 (4-5); 3, 4, outer lateral and upper views of SUI 33657 (4-1); 5-7, upper, outer lateral, and lower views of SUI 33658 (4-3); 8-10, outer lateral, upper, and lower views of SUI 33659 (4-6); 24, 26, 27, outer lateral, upper, and lower views of SUI 33660 (4-3).

11-13, 19-23, 25, 28-32—*Gnathodus bilineatus* (Roundy): 11, 22, upper views of morphotype γ, SUI 33665 (11-33) and SUI 33669 (11-23), respectively; 12, 25, 30, upper views of *morphotype* α, SUI 33666 (11-33), SUI 33671 (11-32), and SUI 33673 (9-8), respectively; 13, 23, upper views of *morphotype* β, SUI 33667 (11-33) and SUI 33670 (7-2), respectively; 19-21, outer lateral, upper, and lower views of *morphotype* δ, SUI 33668 (11-33); 28, 29, 32, outer lateral, lower, and upper views of *morphotype* γ, SUI 33672 (11-23); 31, upper view of *morphotype* δ, SUI 33674 (11-24).

14-18—*Adetognathus unicornis* (Rexroad and Burton): 14, 15, inner lateral and upper views of SUI 33628 (11-33); 16-18, inner lateral, upper, and lower views of SUI 33629 (9-19).

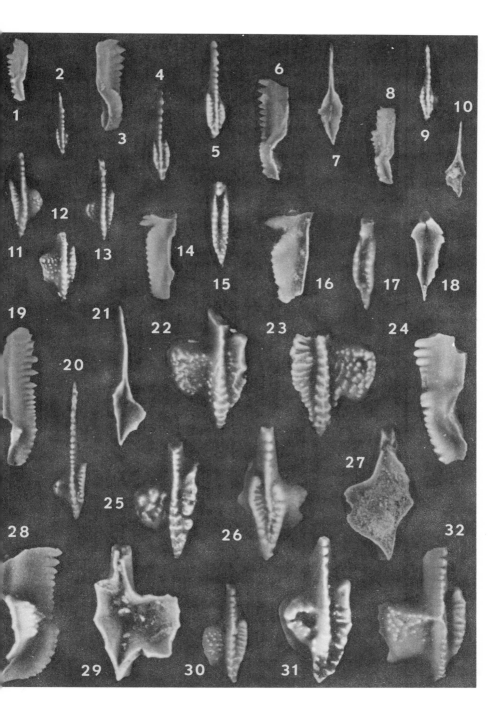

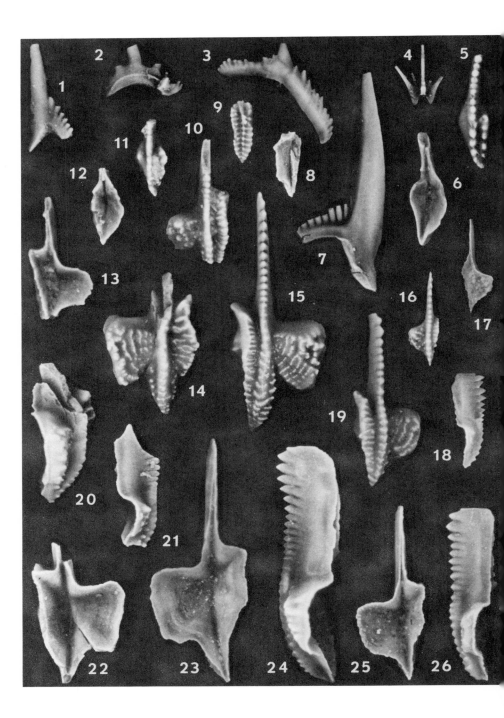

FIGURE 34. *NEOPRIONIODUS, MAGNILATERELLA, OZARKODINA, GNATHODUS,* AND *CAVUSGNATHUS*

Figured specimens 1, 2, 4, and 7 are from the Goddard Formation; 3 is from the Primrose Member of the Golf Course Formation. All others are refigured types. All illustrations are unretouched photographs and are × 31.

1—*Neoprioniodus singularis* (Hass): inner lateral view of SUI 33620 (13-4).

2—*Magnilaterella robusta* Rexroad and Collinson: posterior view of SUI 33621 (7-4).

3—*Ozarkodina* sp.: inner lateral view of SUI 33619 (1-15).

4—*Hibbardella milleri* Rexroad: posterior view of SUI 33623 (11-30).

5, 6—*Gnathodus dilatus* Stauffer and Plummer: upper and lower views of BEG cotype 19150 (=Stauffer and Plummer, 1932, Pl. IV, fig. 10).

7—*Neoprioniodus scitulus* (Branson and Mehl): inner lateral view of SUI 33622 (11-32).

8, 9—*Cavusgnathus nodulifera* Ellison and Graves: lower and upper views of MSM paratype 7164 (=Ellison and Graves, 1941, Pl. 3, fig. 6).

11, 12—*Gnathodus roundyi* Gunnell: upper and lower views of UM cotype 478-3 (=Gunnell, 1931, Pl. 29, figs. 19, 20).

10, 13-26—*Gnathodus bilineatus* (Roundy): 10, 13, 21, upper, lower, and outer lateral views of *morphotype* α, USNM 115101 (=*Polygnathus bilineatus* holotype, refigured from Roundy, 1926, Pl. III, figs. 10a-10c, and from Hass, 1953, Pl. 14, fig. 26); 14, 20, 22, upper, outer lateral, and lower views of *morphotype* β, USNM 115103 (=*Polygnathus texana* holotype, refigured from Roundy, 1926, Pl. III, figs. 13a, 13b, and refigured from Hass, 1953, Pl. 14, fig. 28); 15, 23, 24, upper, lower, and outer lateral views of *morphotype* β, USNM 115104 (refigured hypotype of Hass, 1953, Pl. 14, fig. 29); 16-18, upper, lower, and outer lateral views of *morphotype* α, USNM 115100 (refigured hypotype of Hass, 1953, Pl. 14, fig. 25); 19, 25, 26, upper, lower, and outer lateral views of *morphotype* β, USNM 115102 (refigured hypotype of Hass, 1953, Pl. 14, fig. 27).

FIGURE 35. *IDIOGNATHOIDES* AND *RHACHISTOGNATHUS*

Figured specimens 1-3 are from the Lake Ardmore Member of the Springer Formation; 4-9 and 13-15 are from the Primrose Member of the Golf Course Formation; all others are from the Target Lentil of the Springer Formation. All illustrations are unretouched photographs and are × 31.

1-15—*Idiognathoides noduliferus* (Ellison and Graves): 1-3, outer lateral, lower, and upper views of member of Morphologic Transition Series F, SUI 33612 (3-16); 4-6, outer lateral, upper, and lower views of SUI 33613 (3-4); 7-9, upper, lower, and outer lateral views of a member of Morphologic Transition Series F, SUI 33614 (3-4); 10-12, outer lateral, lower, and upper views of a member of Morphologic Transition Series E, SUI 33611 (5-1); 13-15, upper, lower, and outer lateral views of a member of Morphologic Transition Series E, SUI 33615 (3-4).

16, 17, 24, 30, 31—*Rhachistognathus muricatus* (Dunn): 16, 17, lower and upper views of SUI 33605 (5-1); 24, 30, 31, lower, upper, and outer lateral views of SUI 33607 (5-1).

18-23, 25-29, 32-40—*Rhachistognathus primus* Dunn: 18, 25, 26, upper, oblique outer lateral, and lower views of SUI 33610 (5-1); 19, 23, 35, outer lateral, upper, and lower views of SUI 33609 (5-1); 20, 22, 32, lower, outer lateral, and upper views of SUI 33602 (5-1); 21, upper view of SUI 33603 (5-1); 27, 38, 39, lower, outer lateral, and upper views of SUI 33606 (5-1); 28, 29, 37, outer lateral, upper, and lower views of SUI 33608 (5-1); 33, 34, 36, upper, lower, and outer lateral views of SUI 33604 (5-1); 40, upper view of SUI 33601 (5-1).

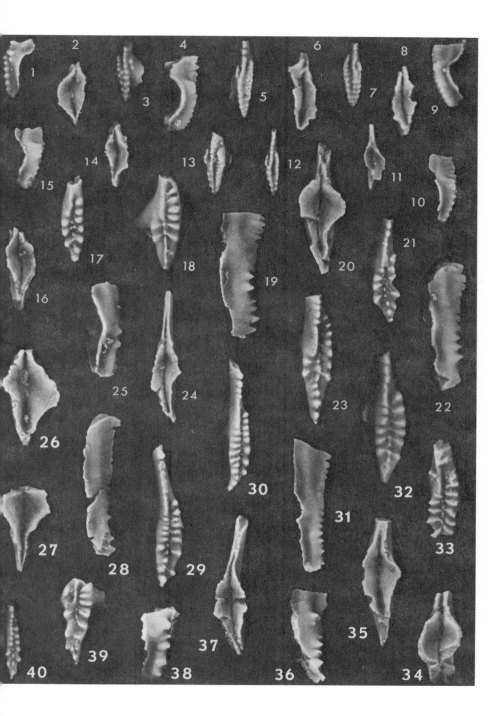

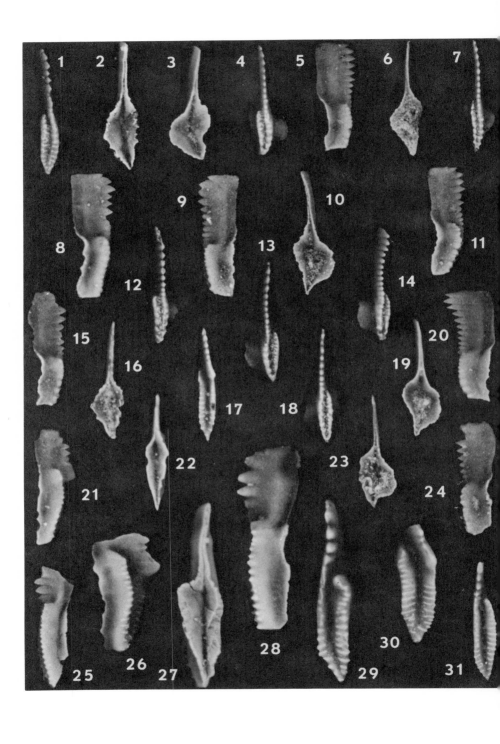

FIGURE 36. *IDIOGNATHOIDES* AND *ADETOGNATHUS*

Figured specimens 1-16 and 18-24 are from the Primrose Sandstone Member of the Golf Course Formation; 17, 21, and 22 are from the Lake Ardmore Member of the Springer Formation; 25-31 are from the Target Lentil of the Springer Formation. All illustrations are unretouched photographs and are × 31.

1-16, 18-20, 23, 24—*Idiognathoides sulcatus sulcatus* Higgins and Bouckaert: 1, 2, 8, upper, lower, and inner lateral views of a member of Morphologic Transition Series F, SUI 33634 (1-16); 3-5, lower and upper and inner lateral views of SUI hypotype 33635 (1-16); 6, 7, 11, lower, upper, and inner lateral views of a member of Morphologic Transition Series F, SUI 33636 (1-16); 9, 10, 13, inner lateral, lower, and upper views of SUI hypotype 33637 (1-16); 12, 15, 16, upper, inner lateral, and lower views of a member of Morphologic Transition Series I, SUI 33638 (1-16); 14, 19, 20, upper, lower, and inner lateral views of SUI 33639 (1-16); 18, 23, 24, upper, lower, and inner lateral views of SUI 33640 (1-16).

17, 21, 22, 25-31—*Adetognathus lautus* (Gunnell): 17, 21, 22, upper, inner lateral, and lower views of right element, SUI 33633 (3-14); 25, 31, inner lateral and upper views of left element, SUI 33630 (5-1); 26, 30, inner lateral and upper views of right element, SUI 33632 (5-1); 27-29, lower, inner lateral, and upper views of left element, SUI 33631 (5-1).

FIGURE 37. *GNATHODUS, IDIOGNATHODUS, IDIOGNATHOIDES, NEOGNATHODUS,* AND *STREPTOGNATHODUS*

Figured specimens 1-9 are from the Goddard Formation; all others are from the Primrose Member of the Golf Course Formation. All illustrations are unretouched photographs and are × 31.

1-9—*Gnathodus commutatus commutatus* (Branson and Mehl): 1, 2, upper and outer lateral views of SUI 33624 (7-4); 3, 4, upper and inner lateral views of SUI 33625 (7-2); 5-7, upper, lower, and outer lateral views of SUI 33626 (11-33); 8, 9, upper and inner lateral views of SUI 33627 (11-33).

10-13, 21—*Idiognathodus sinuosis* Ellison and Graves: 10, 11, 21, upper, lower, and inner lateral views of right element, SUI 33642 (2-2); 12, 13, upper and inner lateral views of left element, SUI 33641 (2-2).

14, 15, 18, 20, 23-26, 36—*Idiognathoides sinuatus* Harris and Hollingsworth: 14, 15, 20, lower, outer lateral, and upper views of right element, SUI 33646 (1-16); 18, 23, 24, upper, lower, and inner lateral views of left element, SUI 33643 (1-11); 25, 26, upper views of left elements, SUI 33644 (1-15) and SUI 33645 (1-6), respectively; 36, upper view of right element, SUI 33647 (1-10).

16, 17, 19—*Neognathodus bassleri bassleri* (Harris and Hollingsworth): outer lateral, lower, and upper views of SUI 33655 (1-10).

22, 31, 32, 37-39—*Neognathodus bassleri symmetricus* (Lane): 22, 31, 32, upper views of SUI 33651 (1-15), SUI 33654 (1-15), and SUI 33653 (1-15), respectively; 37-39, lower, outer lateral, and upper views of SUI 33652 (1-15).

27-30—*Neognathodus* sp.: 27, upper view of SUI 33648 (1-6); 28-30, inner lateral, upper, and lower views of SUI 33649 (1-15).

33-35—*Streptognathodus* spp.: upper, inner lateral, and lower views of SUI 33650 (2-2).

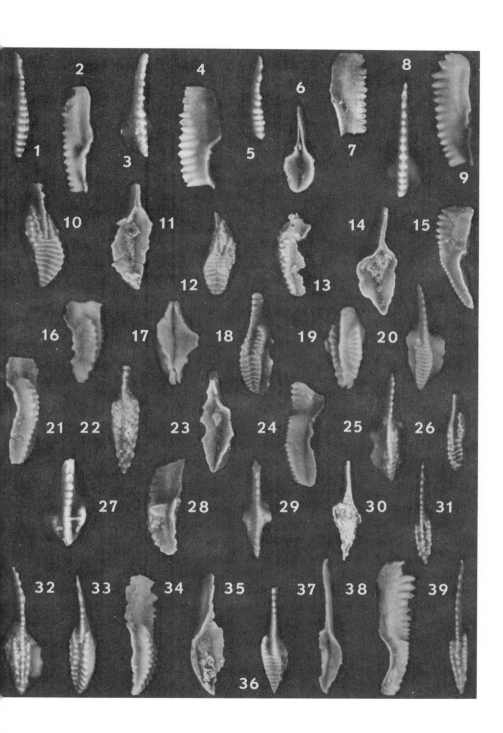

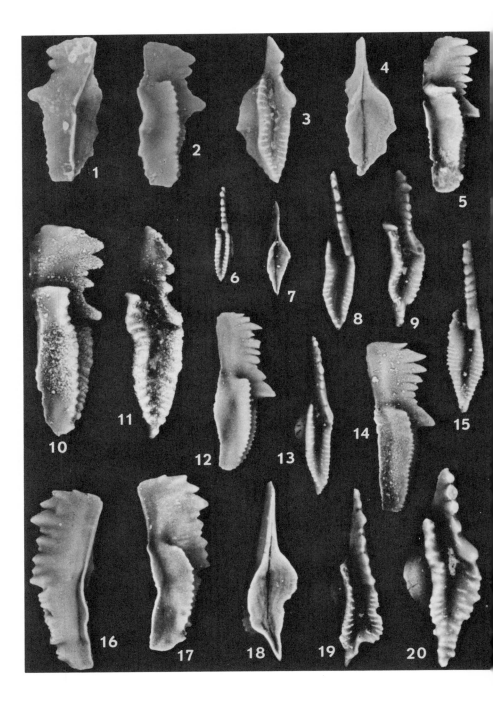

FIGURE 38. *ADETOGNATHUS*

All figured specimens are from the Morrowan of northwestern Arkansas and northeastern Oklahoma, except figures 1-4 which represent a type specimen. All illustrations are unretouched photographs, and 1-4 are × 28 and 5-20 are × 31.

1-4, 6-8, 10-15, 20—*Adetognathus lautus* (Gunnell): all specimens are right elements of *Adetognathus lautus*; 1-4, outer lateral, inner lateral, upper, and lower views, respectively, of UM holotype C515-5 (=Gunnell, 1933, Pl. 33, figs. 7, 8); 6, 7, upper and lower views, respectively, of SUI hypotype 33712 (M-51-14); 8, 20, upper views of SUI hypotypes 33713 (M-64-8), 33718 (M-5-13), respectively; 10, 11, inner lateral and upper views, respectively, of SUI hypotype 33714 (M-42-12); 12, 13, inner lateral and upper views, respectively, of SUI hypotype 33715 (M-26-11); 14, 15, inner lateral and upper views, respectively, of SUI hypotype 33716 (20-8).

5, 9, 16-19—*Adetognathus spathus* (Dunn): 5, 9, inner lateral and upper views, respectively, of SUI hypotype 33711 (M-64-8); 16-19, outer lateral, inner lateral, lower, and upper views of SUI hypotype 33717 (18-2).

FIGURE 39. *IDIOGNATHOIDES, ADETOGNATHUS,* AND *NEOGNATHODUS*

All figured specimens are from the Morrowan of northwestern Arkansas and northeastern Oklahoma. All illustrations are unretouched photographs and are × 31.

1-10—*Idiognathoides sulcatus parvus* Higgins and Bouckaert: 1-4, inner lateral, lower, outer lateral, and upper views, respectively, of SUI hypotype 33725 (20-6); 5-7, 10, upper views of SUI hypotypes 33726-33728, 33730 (all 20-5); 8, 9, lower and upper views, respectively, of SUI hypotype 33729 (20-6).

11-13—*Idiognathoides* n. sp. Harris and Hollingsworth: 11, 12, lower and upper views, respectively, of SUI hypotype 33731 (23-26); 13, upper view of SUI hypotype 33732 (23-34).

14, 15, 19, 20—*Adetognathus lautus* (Gunnell): both are left elements of *Adetognathus lautus*; 14, 15, outer lateral and upper views, respectively, of SUI hypotype 33719 (M-64-8); 19, 20, upper and lower views, respectively, of SUI hypotype 33720 (M-42-7).

16-18, 21-24—*Neognathodus bassleri symmetricus* (Lane): 16, 17, upper and inner lateral views, respectively, of SUI hypotype 33733 (20-16); 18, 21, upper views of SUI hypotypes 33734 (M-51-17), 33735 (M-5-2), respectively; 22-24, upper, outer lateral, and lower views, respectively, of SUI hypotype 33736 (M-64-12).

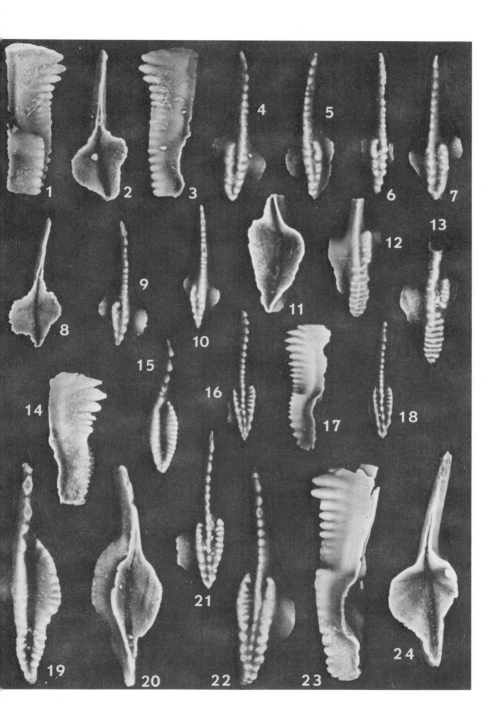

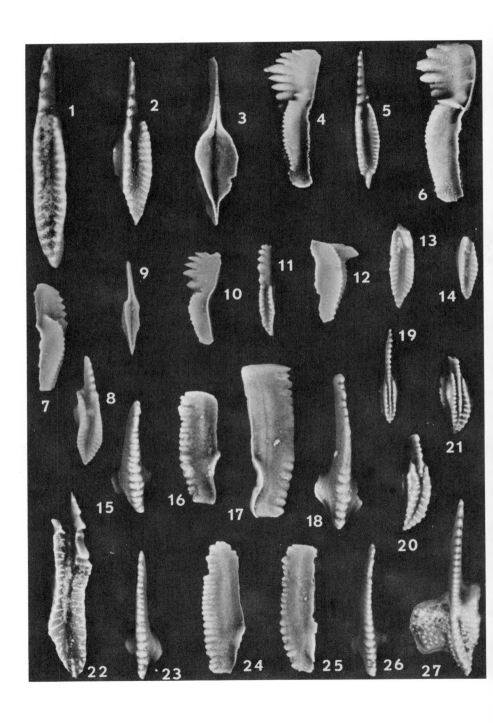

FIGURE 40. *ADETOGNATHUS, SPATHOGNATHODUS, STREPTOGNATHODUS, CAVUSGNATHUS,* AND *GNATHODUS*

Figured specimens 1-6 and 19-21 are from the Morrowan of northwestern Arkansas and northeastern Oklahoma. All others are refigured types. All illustrations are unretouched photographs, and 1-6 and 15-27 are × 31, and 7-14 are × 28.

1-3, 7-14—*Adetognathus lautus* (Gunnell): all specimens are lefts except figured specimens 7, 8, 12, 13; 1, upper view of SUI hypotype 33721 (M-42-17); 2, 3, upper and lower views, respectively, of SUI hypotype 33722 (20-7); 7, 8, inner lateral and upper views, respectively, of hypotype UM C143-4 [= *Adetognathus gigantus* (Gunnell) Ellison, 1941, Pl. 21, fig. 49]; 9-11, lower, inner lateral, and upper views, respectively, of UM lectotype (C502-4, not C502-3 as stated in Ellison, 1941, p. 126; (=Gunnell, 1933, Pl. 31, figs. 67, 68); 12, 13, inner lateral and upper views, respectively, of UM cotype 516-2 [= *Adetognathus missouriensis* (Gunnell), 1933, Pl. 33, fig. 10]; 14, upper view of UM cotype 516-3 [= *Adetognathus missouriensis* (Gunnell), 1933, Pl. 33, fig. 11].

4-6—*Adetognathus spathus* (Dunn): 4, 5, inner lateral and upper views, respectively, of SUI hypotype 33723 (M-26-1); 6, lateral view of SUI hypotype 33724 (M-64-8).

15-18, 23-26—*Spathognathodus commutatus* Branson and Mehl: all syntypes of *Spathognathodus commutatus* are reposited under UM C552-1 (=Branson and Mehl, 1941b, Pl. 19, figs. 1-4); 15, 16, upper and inner lateral views, respectively, of UM cotype; 17, 18, inner lateral and upper views, respectively, of UM cotype; 23, 25, upper and inner lateral views, respectively, of UM cotype; 24, 26, inner lateral and upper views, respectively, of UM cotype.

19-21—*Streptognathodus* spp.: upper views of SUI hypotypes 33737 (M-42-20), 33738 (M-26-14), 33739 (M-26-21), respectively.

22—*Cavusgnathus cristatus* Branson and Mehl: upper view of UM holotype C542-4 (=Branson and Mehl, 1941a, Pl. V, fig. 29).

27—*Gnathodus pustulosus* Branson and Mehl (= *Gnathodus bilineatus* morphotype α, herein): upper view of UM holotype C542-1 (=Branson and Mehl, 1941a, Pl. V, fig. 36).

FIGURE 41. *IDIOGNATHOIDES*

Figured specimens 9-12 and 15-19 are reillustrated types. All other figured specimens are from the Morrowan of northwestern Arkansas and northeastern Oklahoma. All illustrations are unretouched photographs, and 1-8, 13, 14, 18, 20-27 are × 31, and 9-12, 15-17, and 19 are × 28.

1-14, 20-27—*Idiognathoides sinuatus* Harris and Hollingsworth: 1-8, 13, 14, illustrate right elements and 20-27, left elements; 1-3, outer lateral, lower, and upper views, respectively, of SUI hypotype 33740 (20-6); 4-8, upper views of SUI hypotypes 33741 (M-42-18), 33742 (23-46), 33743 (M-5-10), 33744 (20-5), 33745 (20-6), respectively; 9, upper view of form-species *I. corrugatus* USNM lectotype 114715 (=Harris and Hollingsworth, 1933, Pl. 1, figs. 8a, 8b); 10-12, upper views of form-species *I. corrugatus* USNM paralectotypes, all 114714 (fig. 11 = Harris and Hollingsworth, 1933, Pl. 1, fig. 7, and figured specimens 10, 12, were not illustrated by Harris and Hollingsworth, 1933); 13, 14, lower and upper views, respectively, of SUI hypotype 33746 (20-6); 20, 21, 24-27, upper views of SUI hypotypes 33747 (M-42-18), 33748 (24-20), 33750 (21-2), 33751 (20-6), 33752 (M-26-9), 33753 (20-6); 22-23, upper and lower views, respectively, of SUI hypotype 33749 (23-46).

15-17—*Idiognathoides noduliferus* (Ellison and Graves): upper, inner lateral, and lower views of MSM holotype 7143 (=Ellison and Graves, 1941, Pl. 3, fig. 4).

18—*Idiognathoides fossatus* (Branson and Mehl): upper view of UM holotype C60-2 (=Branson and Mehl, 1941b, Pl. 19, figs. 27, 28).

19—*Idiognathoides attenuatus* (Harris and Hollingsworth): upper view of USNM holotype 114712 (=Harris and Hollingsworth, 1933, Pl. 1, figs. 9a, 9b).

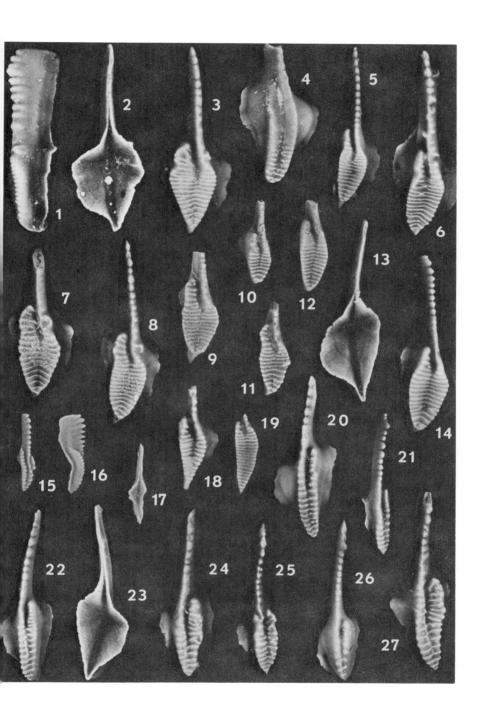

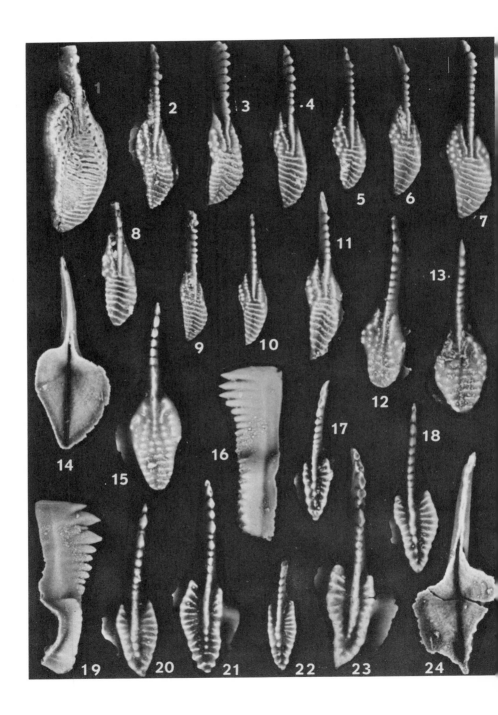

FIGURE 42. *IDIOGNATHODUS* AND *NEOGNATHODUS*

All figured specimens are from the Morrowan of northwestern Arkansas and northeastern Oklahoma. All illustrations are unretouched photographs and are × 31.

1-11—*Idiognathodus sinuosis* Ellison and Graves: all specimens illustrated are right elements of *Idiognathodus sinuosis*; upper views of SUI hypotypes 33754 (M-5-13), 33755 (M-26-15), 33756 (M-5-13), 33757 (23-45), 33758 (M-28-17), 33759 (M-5-13), 33760 (M-5-13), 33761 (21-17), 33762 (M-26-14), 33763 (M-26-14), 33764 (M-28-19), respectively.

12-16—*Idiognathodus klapperi* n. sp.: 12, 13, upper views of SUI paratypes 33765 (23-50), 33766 (23-51), respectively; 14-16, lower, upper, and inner lateral views, respectively, of SUI holotype 33767 (23-51).

17-24—*Neognathodus bassleri bassleri* (Harris and Hollingsworth): 17, 18, 21, 22, upper views of SUI hypotypes 33768 (M-26-11), 33769 (M-5-10), 33771 (M-5-10), 33772 (23-43); 19, 20, outer lateral and upper views, respectively, of SUI hypotype 33770 (20-6); 23, 24, upper and lower views, respectively, of SUI hypotype 33773 (M-5-10).

FIGURE 43. *IDIOGNATHODUS* AND *STREPTOGNATHODUS*

All figured specimens are from the Morrowan of northwestern Arkansas and northeastern Oklahoma. All illustrations are unretouched photographs and are × 31.

1-8, 10-15, 19, 20—*Idiognathodus sinuosis* Ellison and Graves: all figured specimens are left elements of *Idiognathodus sinuosis*; 1, 7, 15, outer lateral, upper, and lower views, respectively, of SUI hypotype 33774 (M-28-19); 2, 3, upper and lower views of SUI hypotype 33775 (23-51); 4-6, 8, 10, 11, 14, 19, 20, upper views of SUI hypotypes 33776 (23-45), 33777 (23-56), 33778 (M-42-18), 33779 (M-28-17), 33780 (M-5-13), 33781 (M-26-14), 33783 (M-42-18), 33784 (M-26-14), 33785 (M-5-13), respectively; 12, 13, outer lateral and upper views, respectively, of SUI hypotype 33782 (M-26-14).

9, 16-18, 21-26—*Streptognathodus expansus* Igo and Koike: 9, 17, 18, inner lateral, upper, and lower views, respectively, of SUI hypotype 33787 (M-28-19); 16, 24, upper views of SUI hypotypes 33788 (M-26-14), 33790 (M-26-14); 21-23, upper, lower, and outer lateral views, respectively, of SUI hypotype 33789 (M-26-15); 25, 26, upper and lower views, respectively, of SUI hypotype 33791 (21-17).

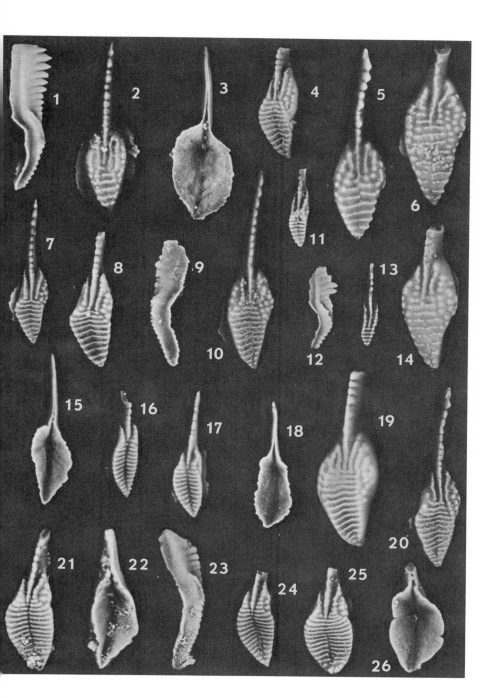

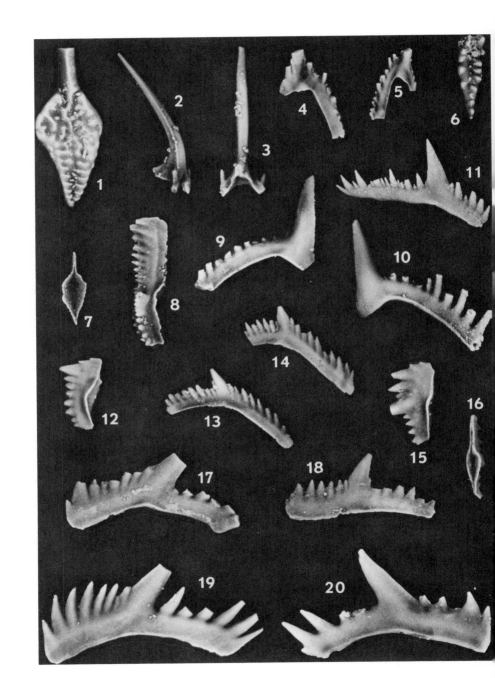

FIGURE 44. *IDIOGNATHODUS, HIBBARDELLA, SYNPRIONIODINA, RHACHISTOGNATHUS, SPATHOGNATHODUS, IDIOGNATHOIDES, NEOPRIONIODUS, HINDEODELLA,* AND *OZARKODINA*

All figured specimens are from the Morrowan of northwestern Arkansas and northeastern Oklahoma. All illustrations are unretouched photographs and are × 31.

1—*Idiognathodus* cf. *I. klapperi* n. sp.: the figured specimen is a right element; upper view of SUI hypotype 33786 (M-42-10).

2, 3—*Hibbardella* sp.: lateral and posterior views, respectively, of SUI hypotype 33792 (20-17).

4, 5—*Synprioniodina* sp.: lateral views of SUI hypotypes 33793 (20-7), 33794 (23-15), respectively.

6—*Rhachistognathus primus* Dunn: upper view of SUI hypotype 33795 (17-2).

7, 12—*Spathognathodus minutus* (Ellison): lower and lateral views, respectively, of SUI hypotype 33796 (20-9).

8—*Idiognathoides macer* (Wirth): inner lateral view of SUI hypotype 33676 (23-31).

9, 10—*Neoprioniodus* sp.: lateral views of SUI hypotypes 33797 (23-17), 33798 (24-21), respectively,

11—*Hindeodella* sp.: lateral view of SUI hypotype 33799 (24-21).

13, 14—*Ozarkodina delicatula* (Stauffer and Plummer): lateral views of SUI hypotypes 33800 (23-10), 33801 (23-16).

15, 16—*Spathognathodus* n. sp.: lateral and lower views, respectively, of SUI hypotype 33802 (20-7).

17-20—*Ozarkodina* spp.: lateral views of SUI hypotypes 33803 (23-36a), 33804 (20-1), 33805 (24-23), 33806 (24-20), respectively.

References Cited

Bennison, A. P., 1954, Target Limestone, a new member of Springer Formation, Carter County, Oklahoma: Am. Assoc. Petroleum Geologists Bull., v. 38, p. 913-914.

Bischoff, G., 1957, Die Conodonten-Stratigraphie des rheno-herzynischen Unterkarbons mit Berücksichtigung der Wocklumeria-Stufe und der Devon/Karbon-Grenze: Abh. Hess. Landesamt Bodenforsch., Heft 19, 64 p.

Bouckaert, J., and Higgins, A. C., 1970, The position of the Mississippian-Pennsylvanian Boundary in the Namurian of Belgium: Colloque sur la stratigraphie du Carbonifère, 8ᵉ, Université de Liége, 1970, Comptes Rendus, v. 55, p. 197-204.

Branson, C. C., 1959, Regional relationship of Ouachita Mississippian and Pennsylvanian rocks, in Cline, L. M., and others, eds., The geology of the Ouachita Mountains—A symposium: Dallas-Ardmore Geol. Socs., p. 118-121.

Branson, E. B., and Mehl, M. G., 1933, Conodonts from the Bainbridge (Silurian) of Missouri: Missouri Univ. Studies, Conodont Studies No. 1, v. VIII, p. 39-52, Pl. 3.

_____1941a, Caney conodonts of Upper Mississippian age: Denison Univ. Sci. Lab. Jour., v. 35, art. 5, p. 167-178, Pl. V (date of imprint, 1940).

_____1941b, New and little known Carboniferous conodont genera: Jour. Paleontology, v. 15, p. 97-106, Pl. 19.

Cheney, M. R., Dott, R. H., Hake, B. F., Moore, R. C., Newell, N. D., Thomas, H. D., and Tomlinson, C. W., 1945, A classification of Mississippian and Pennsylvanian rocks of North America: Am. Assoc. Petroleum Geologists Bull., v. 29, p. 125-163.

Clarke, W. J., 1960, Scottish Carboniferous conodonts: Edinburgh Geol. Soc. Trans., v. 18, pt. 1, 31 p.

Collinson, C., Scott, A. J., and Rexroad, C. B., 1962, Six charts showing biostratigraphic zones and correlations based on conodonts from the Devonian and Mississippian rocks of the upper Mississippi Valley: Illinois Geol. Survey Circ. 328, 32 p.

Collinson, C., Rexroad, C. B., and Thompson, T. L., 1971, Conodont zonation of the North American Mississippian, in Sweet, W. C., and Bergström, S. M., eds., Symposium on conodont biostratigraphy: Geol. Soc. America Mem. 127, p. 353-394.

Cooper, C. L., 1947, Upper Kinkaid (Mississippian) microfauna from Johnson County, Illinois: Jour. Paleontology, v. 21, p. 81-94, Pls. 19-23.

Dott, R. H., 1941, Regional stratigraphy of the mid-continent: Am. Assoc. Petroleum Geologists Bull., v. 25, p. 1619-1705.

Dunn, D. L., 1965, Late Mississippian conodonts from the Bird Spring Formation in Nevada: Jour. Paleontology, v. 39, p. 1145-1150, Pl. 140.

_____1966, New Pennsylvanian conodonts from southwestern United States: Jour. Paleontology, v. 40, p. 1294-1303, Pls. 157, 158.

_____1970a, Middle Carboniferous conodonts from western United States and phylogeny of the platform group: Jour. Paleontology, v. 44, p. 312-342, Pls. 61-64.

_____1970b, Conodont zonation near the Mississippian-Pennsylvanian boundary in western United States: Geol. Soc. America Bull., v. 81, p. 2959-2974.

_____1971, Considerations of the *Idiognathoides-Declinognathodus-Neognathodus* complex of Middle Carboniferous conodonts: Lethaia, v. 4, p. 15-19.

Easton, W. H., 1942, The Pitkin Limestone of northern Arkansas: Arkansas Geol. Survey Bull. 8, 115 p.

Elias, M. K., 1956a, Upper Mississippian and Lower Pennsylvanian formations of south-central Oklahoma, *in* Hicks, I. C. and others, eds., Petroleum geology of southern Oklahoma—A symposium: Ardmore Geol. Soc., v. 1, p. 56-134.

_____1956b, Upper Mississippian and Lower Pennsylvanian formations of southern Oklahoma— Corrections: Am. Assoc. Petroleum Geologists Bull., v. 40, p. 2513.

_____1959, Some Mississippian conodonts from the Ouachita Mountains, *in* Cline, L. M., and others, eds., The geology of the Ouachita Mountains, Symposium: Dallas-Ardmore Geol. Socs., p. 141-165, Pls. 1, 2.

_____1960, Marine Carboniferous of North America and Europe: Cong. Avanc. Études Stratig. et Geol. Carbonifère, 4th, Comptes Rendus, t. 1, p. 151-161.

Ellison, S., 1941, Revision of the Pennsylvanian conodonts: Jour. Paleontology, v. 15, p. 107-143, Pls. 20-23.

Ellison, S., and Graves, R. W., Jr., 1941, Lower Pennsylvanian (Dimple Limestone) conodonts of the Marathon region, Texas: Missouri Univ., School of Mines and Metallurgy Bull., Tech. Serv., v. 14, no. 3, 21 p.

Fay, R. O., 1952, Catalogue of conodonts: Kansas Univ. Paleont. Contr., Vertebrata, art. 3, 206 p.

Felix, C. J., and Burbridge, P. P., 1967, Palynology of the Springer Formation of southern Oklahoma, U.S.A.: Palaeontology, v. 10, p. 349-425, Pls. 53-66.

Furnish, W. M., Quinn, J. H., and McCaleb, J. A., 1964, The Upper Mississippian ammonoid *Delepinoceras* in America: Palaeontology, v. 7, p. 173-180, Pl. 30.

Girty, G. H., and Roundy, P. V., 1923, Notes on the Glenn Formation of Oklahoma with consideration of new paleontologic evidence: Am. Assoc. Petroleum Geologists Bull., v. 7, p. 331-347.

Globensky, Y., 1967, Middle and Upper Mississippian conodonts from the Windsor Group of the Atlantic Provinces, Canada: Jour. Paleontology, v. 41, p. 432-448, Pls. 55-58.

Goldston, W. L., Jr., 1922, Differentiation and structure of the Glenn Formation: Am. Assoc. Petroleum Geologists Bull., v. 4, p. 1-23.

Gordon, M., Jr., 1965, Carboniferous cephalopods of Arkansas: U.S. Geol. Survey Prof. Paper 460, 322 p.

Gould, C. N., and Wilson, R. A., 1927, The upper Paleozoic rocks of Oklahoma: Oklahoma Geol. Survey Bull. 41, 66 p.

Gunnell, F. H., 1931, Conodonts from the Fort Scott Limestone of Missouri: Jour. Paleontology, v. 5, p. 244-252, Pl. 29.

_____1933, Conodonts and fish remains from the Cherokee, Kansas City, and Wabaunsee Groups of Missouri and Kansas: Jour. Paleontology, v. 7, p. 261-297, Pls. 31-33.

Harlton, B. H., 1933, Micropaleontology of the Pennsylvanian Johns Valley Shale of the Ouachita Mountains, Oklahoma, and its relationship to the Mississippian Caney Shale: Jour. Paleontology, v. 7, p. 3-29, Pls. 1-7.

_____1934, Carboniferous stratigraphy of the Ouachitas with special study of the Bendian: Am. Assoc. Petroleum Geologists Bull., v. 18, p. 1018-1049.

_____1938, Stratigraphy of the Bendian of the Oklahoma salient of the Ouachita Mountains: Am. Assoc. Petroleum Geologists Bull., v. 22, p. 852-914.

_____1956, The Harrisburg Trough, Stephens and Carter Counties, Oklahoma, in Hicks, I. C., and others, eds., Petroleum geology of southern Oklahoma—A symposium: Ardmore Geol. Soc., v. 1, p. 135-143.

Harris, R. W., and Hollingsworth, R. V., 1933, New Pennsylvanian conodonts from Oklahoma: Am. Jour. Sci., v. 25, p. 193-204.

Hass, W. H., 1951, Age of Arkansas Novaculite: Am. Assoc. Petroleum Geologists Bull., v. 35, p. 2526-2541.

_____1953, Conodonts of the Barnett Formation of Texas: U.S. Geol. Survey Prof. Paper 243-F, p. 69-94, Pls. 14-16, Fig. 4.

_____1959, Conodonts from the Chappel Limestone of Texas: U.S. Geol. Survey Prof. Paper 294-J, p. 365-399, Pls. 46-50.

_____1962, Conodonts, in Treatise on invertebrate paleontology, Pt. W, Miscellanea: Lawrence, Kansas, Geol. Soc. America and Kansas Univ. Press, p. W3-W69.

Henbest, L. G., 1953, Morrow Group and lower Atoka Formation of Arkansas: Am. Assoc. Petroleum Geologists Bull., v. 37, p. 1935-1953.

_____1962a, Type sections for the Morrow Series of Pennsylvanian age, and adjacent beds, Washington County, Arkansas: U.S. Geol. Survey Prof. Paper 450-D, art. 130, p. D38-D41.

_____1962b, New members of the Bloyd Formation of Pennsylvanian age, Washington County, Arkansas: U.S. Geol. Survey Prof. Paper 450-D, art. 131, p. D42-D44.

Hicks, I. C., Westheimer, J., Tomlinson, C. W., Putman, D. M., and Selk, E. L., eds., 1956, Petroleum geology of southern Oklahoma—A symposium: Ardmore Geol. Soc., v. 1, p. 2-13.

Higgins, A. C., 1961, Some Namurian conodonts from North Staffordshire: Geol. Mag., v. 98, p. 210-224, Pls. 10-12.

Higgins, A. C., and Bouckaert, J., 1968, Conodont stratigraphy and paleontology of the Namurian of Belgium: Belgique Service Géol. Mem. 10, 64 p.

Hinde, G. J., 1900, in Smith, J., Conodonts from the Carboniferous limestone strata of the west of Scotland: Glasgow Nat. Hist. Soc. Trans., v. 5, n.s., p. 336-346, Pls. 9-10.

Holmes, G. B., 1928, A bibliography of the conodonts with descriptions of Early Mississippian species: U.S. Natl. Mus. Proc., v. 72, art. 5, p. 1-38, Pls. 1-11.

Huffman, G. G., 1958, Geology of the flanks of the Ozark Uplift, northeastern Oklahoma: Oklahoma Geol. Survey Bull. 77, 281 p.

Igo, H., and Koike, T., 1964, Carboniferous conodonts from the Omi Limestone, Niigata Prefecture, central Japan (studies of Asian conodonts, Pt. I): Palaeont. Soc. Japan Trans. and Proc., n.s., no. 53, p. 179-193, Pls. 27, 28.

_____1965, Carboniferous conodonts from Yobara, Akiyoshi Limestone, Japan (studies of Asiatic conodonts, Pt. II): Palaeont. Soc. Japan Trans. and Proc., n.s., no. 59, p. 83-91, Pls. 8, 9.

_____1968, Carboniferous conodonts from Kuantan, Malaya, in Kobayashi, T., and Toriyama, R., eds., Geology and palaeontology of southeast Asia: Tokyo Univ. Press, v. 5, contr. LV, p. 26-30, Pl. III.

Jones, D. J., 1941, The conodont fauna of the Seminole Formation: Chicago Univ. Libraries, 55 p.

Klapper, G., and Ormiston, A. R., 1969, Lower Devonian conodont sequence, Royal Creek, Yukon Territory and Devon Island, Canada, with a section on Devon Island stratigraphy: Jour. Paleontology, v. 43, p. 1-27, Pls. 1-6.

Koike, T., 1967, A Carboniferous succession of conodont faunas from the Atetsu Limestone in southwest Japan (studies of Asiatic conodonts, Pt. VI): Tokyo Kyoiku Daigaku Sci. Repts., Sec. c, Geology, Mineralogy, and Geography, v. 9, no. 93, p. 279-318.

Lane, H. R., 1967, Uppermost Mississippian and Lower Pennsylvanian conodonts from the type Morrowan region, Arkansas: Jour. Paleontology, v. 41, p. 920-942, Pls. 119-123.

——1968, Symmetry in conodont element-pairs: Jour. Paleontology, v. 42, p. 1258-1263.

Lane, H. R., Merrill, G. K., Straka, J. J., II, and Webster, G. D., 1971, North American Pennsylvanian conodont biostratigraphy, in Sweet, W. C., and Bergström, S. M., eds., Symposium on conodont biostratigraphy: Geol. Soc. America Mem. 127, p. 395-414.

Lane, H. R., Sanderson, G. A., and Verville, G. J., 1972, Uppermost Mississippian-basal Middle Pennsylvanian conodonts and fusulinids from several exposures in the south-central and southwestern United States: Internat. Geol. Cong., 24th, Montreal 1972, Sec. 7, Paleontology, p. 549-555.

Lang, R. C., III, 1957, The Criner Hills: A key to the geologic history of southern Oklahoma: Ardmore Geol. Soc. Guidebook, Criner Hills Field Conf., p. 18-25.

——1966, The Pennsylvanian rocks of the Lake Murray area, in Pennsylvanian of the Ardmore Basin, southern Oklahoma: Ardmore Geol. Soc. Guidebook, 14th Field Conf., p. 13-18.

Laudon, R. B., 1958, Chesterian and Morrowan rocks in the McAlester Basin of Oklahoma: Oklahoma Geol. Survey Circ. 46, 30 p.

Lindström, M., 1964, Conodonts: Amsterdam, Elsevier Publishing Co., 196 p.

Mamet, B. L., and Skipp, B. A., 1970, Preliminary foraminiferal correlations of early Carboniferous strata in the North American Cordillera: Colloque sur la stratigraphie du Carbonifère, 8ᵉ, Université de Liége, 1970, Comptes Rendus, v. 55, p. 327-348.

McLaughlin, K. P., 1952, Microfauna of the Pennsylvanian Glen Eyrie Formation, Colorado: Jour. Paleontology, v. 26, p. 613-621, Pls. 82, 83.

Meischner, D., 1970, Conodonten-Chronologie des deutschen Karbons: Cong. Internat. Strat. Géol. Carbonifère, 6th, Sheffield 1967, Comptes Rendus, v. 3, p. 1169-1180.

Moore, R. C., 1934, The origin and age of the boulder-bearing Johns Valley Shale in the Ouachita Mountains of Arkansas and Oklahoma: Am. Jour. Sci., 5th ser., v. 27, p. 432-453.

Moore, R. C., and Thompson, M. L., 1949, Main divisions of Pennsylvanian Period and System: Am. Assoc. Petroleum Geologists Bull., v. 33, p. 275-302.

Moore, R. C., and others, 1944, Correlation of Pennsylvanian formations of North America: Geol. Soc. America Bull., v. 55, p. 657-706.

Müller, K. J., 1962, Zur systematischen Einteilung der Conodontophorida: Paläont. Zeitschr., Bd. 36, p. 109-117.

Murray, F. N., and Chronic, J., 1965, Pennsylvanian conodonts and other fossils from insoluble residues of the Minturn Formation (Desmoinesian), Colorado: Jour. Paleontology, v. 39, p. 594-610, Pls. 71-73.

Palmieri, V., 1969, Upper Carboniferous conodonts from limestones near Murgon, south-east Queensland: Queensland Geol. Survey Pub. 341, Palaeont. Paper no. 17, 13 p.

Pander, C. H., 1856, Monographie der fossilen Fische des Silurischen Systems der Russisch-Baltischen Gouvernements: St. Petersburg, Kaiserl. Akad. Wiss., 91 p.

Peace, H. W., II, 1965, The Springer Group of the southeastern Anadarko Basin in Oklahoma: Shale Shaker, The Digest V, 1968, p. 280-297.

Philip, G. M., 1965, Lower Devonian conodonts from the Tyers area, Gippsland, Victoria: Royal Soc. Victoria Proc., v. 79, pt. 1, p. 95-117, Pls. 8-10.

Purdue, A. H., 1907, Winslow, Arkansas—Indian Territory: U.S. Geol. Survey Geol. Folio 154, 6 p.

Quinn, J. H., 1966, Genus Reticuloceras in America: Oklahoma Geol. Notes, v. 26, no. 1, p. 13-20.

Quinn, J. H., and Saunders, W. B., 1968, The ammonoids Hudsonoceras and Baschkirites in the Morrowan Series of Arkansas: Jour. Paleontology, v. 42, p. 397-402, Pl. 57.

Rexroad, C. B., 1957, Conodonts from the Chester Series in the type area of southwestern Illinois: Illinois Geol. Survey, R.I. 199, 43 p.

———1958, Conodonts from the Glen Dean Formation (Chester) of the Illinois Basin: Illinois Geol. Survey, R.I. 209, 27 p.

Rexroad, C. B., and Burton, R. C., 1961, Conodonts from the Kinkaid Formation (Chester) in Illinois: Jour. Paleontology, v. 35, p. 1143-1158, Pls. 138-141.

Rexroad, C. B., and Collinson, C. W., 1963, Conodonts from the St. Louis (Valmeyeran Series) of Illinois, Indiana, and Missouri: Illinois Geol. Survey Circ. 355, 28 p.

Rexroad, C. B., and Furnish, W. M., 1964, Conodonts from the Pella Formation (Mississippian) south-central Iowa: Jour. Paleontology, v. 38, p. 667-676, Pl. 111.

Rexroad, C. B., and Nicoll, R. S., 1965, Conodonts from the Menard Formation (Chester Series) of the Illinois Basin: Indiana Geol. Survey Bull. 35, 28 p.

Rhodes, F. H. T., Austin, R. L., and Druce, E. C., 1969, British Avonian (Carboniferous) conodont faunas, and their value in local and intercontinental correlation: British Mus. (Nat. History) Bull., Geology, supp. 5, 313 p.

Rich, M., 1961, Stratigraphic section and fusulinids of the Bird Spring Formation near Lee Canyon, Clark County, Nevada: Jour. Paleontology, v. 35, p. 1159-1180, Pls. 142-146.

Roundy, P. V., 1926, in Roundy, P. V., Girty, G. H., and Goldman, M. I., eds., Mississippian formations of San Saba County, Texas; Pt. II, The micro-fauna: U.S. Geol. Survey Prof. Paper 146, p. 5-17, Pls. 1-4.

Sadlick, W., 1955, Carboniferous formations of northeastern Uinta Mountains: Wyoming Geol. Assoc. Guidebook, 10th Ann. Field Conf., p. 49-59.

Sanderson, G. A., and King, W. E., 1964, Paleontological evidence for the age of the Dimple Limestone, in Permian basin section, The filling of the Marathon geosyncline—A symposium and guidebook: Soc. Econ. Paleontologists and Mineralogists, p. 31-34.

Scott, A., and Collinson, C., 1961, Conodont faunas from the Louisiana and McCraney Formations of Illinois, Iowa, and Missouri: Kansas Geol. Soc. Guidebook, 26th Ann. Field Conf., p. 110-141.

Simonds, F. W., 1891, The Geology of Washington County: Arkansas Geol. Survey Ann. Rept. 1888, v. 4, p. 1-148.

Stauffer, C. R., and Plummer, H. J., 1932, Texas Pennsylvanian conodonts and their stratigraphic relations: Texas Univ. Bull., no. 3201, p. 13-50, Pls. 1-4.

Stibane, F. R., 1967, Conodonten des Karbons aus den nördlichen Anden Südamerikas: Neues Jahrb. Geologie u. Paläontologie Abh., Bd. 128, Heft 3, p. 329-340, Pls. 35-37.

Straka, J. J., II, 1972, Conodont evidence of age of Goddard and Springer Formations, Ardmore Basin, Oklahoma: Am. Assoc. Petroleum Geologists Bull., v. 56, p. 1087-1099.

Straka, J. J., II, and Lane, H. R., 1970, Evolution of some Lower Pennsylvanian conodont species: Lethaia, v. 3, p. 41-49.

Sturgeon, M. T., and Youngquist, W., 1949, Allegheny conodonts from eastern Ohio: Jour. Paleontology, v. 23, p. 380-386, Pls. 74, 75.

Sutherland, P. K., and Henry, T. W., 1974, Stratigraphy and biostratigraphy of the Morrow Series in northeastern Oklahoma: Oklahoma Geol. Survey Bull. (in press).

Taff, J. A., 1903, Tishomingo Folio (Indian Territory): U.S. Geol. Survey Geol. Folio 98.

Thompson, T. L., and Goebel, E. D., 1968, Conodonts and stratigraphy of Meramecian Stage (Upper Mississippian) in Kansas: Kansas Geol. Survey Bull. 192, 56 p.

Tomlinson, C. W., 1929, The Pennsylvanian System in the Ardmore Basin: Oklahoma Geol. Survey Bull. 46, 79 p.

———1959, Best exposures of various strata in Ardmore Basin, 1957, in Mayes, J. W., and others, eds., Petroleum geology of southern Oklahoma—A symposium: Ardmore Geol. Soc., v. 2, p. 302-334.

Tomlinson, C. W., and Bennison, A., 1960, Tiff Member of Goddard Formation: Oklahoma Geology Notes, v. 20, no. 5, p. 123-124.

Tomlinson, C. W., and McBee, W., Jr., 1959, Pennsylvanian sediments and orogenies of Ardmore district, Oklahoma, in Mayes, J. W., and others, eds., Petroleum geology of southern Oklahoma—A symposium: Ardmore Geol. Soc., v. 2; p. 3-52.

——1962, Pennsylvanian sediments and orogenies of Ardmore district, Oklahoma, in Branson, C. C., ed., Pennsylvanian System in the United States—A symposium: Am. Assoc. Petroleum Geologists, p. 461-500.

Ulrich, E. O., 1904, in Adams, G. I., Purdue, A. H., and Burchard, E. E., eds., Zinc and lead deposits of northern Arkansas: U.S. Geol. Survey Prof. Paper 24, 118 p.

Voges, A., 1959, Conodonten aus dem Unterkarbon I und II (Gattendorfia- und Pericyclus-Stufe) des Sauerlandes: Paläont. Zeitschr., Bd. 33, p. 266-314, Pls. 33-35.

Weaver, C. E., 1958, Geologic interpretation of argillaceous sediments. Pt. II. Clay petrology of Upper Mississippian-Lower Pennsylvanian sediments of central United States: Am. Assoc. Petroleum Geologists Bull., v. 42, p. 272-309.

Webster, G. D., 1969, Chester through Derry conodonts and stratigraphy of northern Clark and southern Lincoln Counties, Nevada: California Univ. Pubs. Geol. Sci., v. 79, 121 p.

Webster, G. D., and Lane, N. G., 1967, Mississippian-Pennsylvanian Boundary in southern Nevada, in Teichert, C., and Yochelson, E. L., eds., Essays in paleontology and stratigraphy, R. C. Moore Commemorative Volume: Lawrence, Kansas Univ. Press, Dept. Geol. Spec. Pub. 2, p. 503-522.

Weller, J. M., and others, 1948, Correlation of the Mississippian formations of North America: Geol. Soc. America Bull., v. 59, p. 91-196.

Westheimer, J. M., 1956, The Goddard Formation, in Hicks, I. C., and others, eds., Petroleum geology of southern Oklahoma—A symposium: Ardmore Geol. Soc., v. 1, p. 392-396.

Wirth, M., 1967, Zur Gliederung des höheren Paläozoikums (Givet-Namur) im gebiet des Quinto Real (Westpyrenäen) mit hilfe von conodonten: Neues Jahrb. Geologie u. Paläontologie Abh., Bd. 127, Heft 2, p. 179-244, Pls. 19-23.

Youngquist, W., and Downs, R. H., 1949, Additional conodonts from the Pennsylvanian of Iowa: Jour. Paleontology, v. 23, p. 161-171, Pls. 30-31.

Youngquist, W., and Miller, A. K., 1949, Conodonts from the Late Mississippian Pella beds of south-central Iowa: Jour. Paleontology, v. 23, p. 617-622, Pl. 101.

MANUSCRIPT RECEIVED BY THE SOCIETY JUNE 12, 1972

Index